Customer Service Over The Phone

5th Edition

By Stephen Coscia

Telecom Books, 12 West 21 Street New York NY 10010
212-691-8215   1-800-LIBRARY

Published by CMP Books
An Imprint of CMP Media LLC
12 West 21 Street
New York, NY 10010

ISBN 1-57820-046-6

For individual orders, and for information on special discounts for quantity orders, please contact:

CMP Books
6600 Silacci Way
Gilroy, CA 95020
Tel: 800-500-6875 or 408-848-3854
Fax: 408-848-5784
Email: cmp@rushorder.com

Distributed to the book trade in the U.S. and Canada by
Publishers Group West
1700 Fourth St., Berkeley, CA 94710

Manufactured in the United States of America

Dedications

Veronica H. Coscia, my wife and editor, for helping with the final text along with much love and support.

Dan Garrett came up with the term contain, qualify and correct.

Bob Kay, a friend, whose teaching goes beyond the classroom.

Christine Kern and her staff, for all the help and support.

To my friends who provided encouragement, guidance and constructive criticism: Rob Bonanno, Joe Paschall, Alan Blake, Bert Neikirk, Steve Pretti, Phil and Elizabeth Huston, James Dinh, Roy Elkins, Tom Tracy, Vic Adams, Joe Baiocco, Dr. Steven Marcus, Marge Pierce, Debbie Willsey, Dolly Reingold, John Collier, Christine Yannes and Richard Norman.

Acknowledgements

Some text in Chapter 7 appeared in the March 1992 issue of TeleProfessional™.

Cartoon images in figures 1, 2, 3, 13, 14, & 15 are from Clickart® Business Cartoons by T/Maker Company. All rights reserved.

Table of Contents

Preface To The 5th Edition:

In 1990, when I first had the idea for this book, I thought my offering would be too late to make any significant difference in the customer service business. Numerous industry insiders warned me that customer service was a very small niche market and interest would be scarce if at all. In addition, there were already a few good customer service books out there and, many believed that customer service professionals had all the material they needed. Still, I wanted a chance to share *my ideas*.

I guess that no one could have imagined the explosive growth in call centers, during the last few years, that has placed customer service strategies in such high demand. This demand has propelled this book's popularity and caused the publisher to keep running out of stock. This ongoing condition led to phone calls from my publisher, urging me to prepare a subsequent version of this book.

So here is the 5th edition of my original book. I've written a few new stories, added new illustrations and updated text wherever possible. As always, I extend an open invitation to contact me about any of this book's content.

Stephen Coscia
P.O. Box 786
Havertown PA 19083
610-658-4417
coscia@worldnet.att.net
www.coscia.com

Introduction:

The inception of this book was a speech I delivered at a telecommunication conference on the subject of, *Handling Calls from Difficult Customers and the Related Stress.* As I described calls from difficult customers, I observed nods of approval and empathy. Immediately after my presentation, many attendees thanked me for accurately describing the customer service environment and for the practical techniques on problem solving. They told me that information on handling difficult customers and problem solving is rare because most professionals are not willing to admit their companies have problems or that their customers get angry. I knew this was a topic worth covering in more detail.

In this book, I have expanded on the ideas of that speech to make this information available to all customer service representatives (CSRs). This includes receptionists, customer service representatives, technical support representatives, airline reservation representatives, hotel clerks and financial account representatives.

The purpose of this book is to enhance the one-on-one interaction between CSRs and customers. Included is a chapter on the importance of using the correct words when working with customers which will enable you to be proactive, efficient and helpful rather than antagonistic.

There are also a few chapters on problem solving. These chapters express the importance of containing adverse situations prior to resolving them. Included is a simple troubleshooting flowchart which when applied, provides a consistent problem solving method. It will help develop your problem solving skills at a time when customers are more demanding than ever.

The last few chapters describe the social and business changes that have resulted in rising customer expectations and an increase in on the job stress. Within each chapter there are short stories, case studies and practical illustrations. While the characters in my stories are fictitious, real-life experiences inspired each event.

Prior to writing this book, I conducted a survey involving hundreds of CSRs. I asked what subjects would be most beneficial and more than half of those surveyed suggested I focus on the following three issues: (1) How to communicate correct words when working with adversity, (2) How to handle the stress of customer service and (3) How to overcome obstacles and solve problems. These three issues are covered in this book as well as other pertinent customer service information.

Most customers will not bother to complain about bad service. They'll just buy somewhere else the next time. Satisfying customers and solving problems with a sense of urgency is imperative in today's fast paced business climate. Listening to customer's complaints, once considered a drain on resources, has become a marketing strategy for customer retention. Managers looking to implement this new strategy may sense some apprehension among front line people because CSRs have to bear the brunt of customer anger and frustration. The techniques in this book will help CSRs resolve problems, satisfy customer complaints and increase customer retention.

The irate customer is at war with you and your company and he thinks your company fired the first shot. Your job is to remove any resemblance of a conflict by absorbing some rash comments and verbal abuse, by not retaliating with similar behavior and by containing the situation and resolving the problem.

You are in the people business. Whether you know it or not, if you did not like people, you would not be doing what you're doing. Customer service is more than a function, it's an attitude! Effective CSRs are customer driven and realize that their customer's needs are their company's needs. These CSRs do this with a combination of (1) an immediate response to a customer's question, problem or complaint, (2) the proper use of words and (3) a consistent method of problem solving. All three provide a winning formula for enhanced customer satisfaction.

Chapter 1:
Avoiding Obstacles

The Truth in Customer Service

Customer service, when boiled down to its purest form, is the act of one person helping another person. This interaction determines how a company's service will be perceived by customers. In order for this one-on-one relationship to work, both the customer and the CSR must be honest with each other. No problem can be resolved to either party's mutual satisfaction unless the truth about the cause of a problem is revealed.

Almost all customers are honest and straight forward about their reasons for calling for help. It doesn't make sense for customers to invent a reason for calling unless he/she is lonely and just needs someone to talk to (which happens sometimes).

Good business ethics start here. There are no short cuts around the truth and no one will last for long in business without being honest. Customer service is a wonderful position for individuals with motivation, drive and a genuine concern for people. The skills learned in the customer service department will have a positive impact on one's career if you start off by always telling the truth no matter how much it hurts. My first rule of customer service is: It is better to disappoint a customer with the truth than to satisfy a customer with a lie.

Occasionally, the truth appears to be an obstacle because it (1) is uncomfortable, (2) results in more work or (3) uncovers weaknesses in either you or your company. However: (1) uncomfortable situations are temporary and they are wonderful growth opportunities for people with resilient personalities, (2) if doing the right thing means working harder then do it and (3) uncovering weaknesses is a golden opportunity to improve one's service techniques.

When CSRs tell the truth, the customer knows where he stands. The CSR may either offer solutions to the customer's problems or may relay what the company will do to help him resolve his difficulty. Vital to all communications with customers, honesty is the best policy. Without the truth, you and your company have nothing.

The Service Placebo

A placebo is a preparation containing no medicine given to humor a patient or for a psychological effect. Placebo in Latin means, "I will please." Placebo service, likewise, is an imitation of the real thing. To an unsuspecting customer, placebo service seems like the real thing, until your placebo is inevitably exposed. Placebo service does not work. Don't be tempted into thinking that placebo service is your only choice.

Jan Boyle, in the following example, gets herself into trouble because she satisfied her customer with a lie. Jan's experience is not unique. Unfortunately, many CSR's learn what not to do by making the same mistake (hopefully only once). If you are an entry level CSR, I hope this example will prevent you from providing placebo service.

Figure 1

The First Rule of Customer Service

It's better to disappoint a customer with the truth than to satisfy a customer with a lie!

Jan Boyle learned about placebo service the hard way. She was not having a good day. Jan thought to herself, "There must be a full moon." Her customers were asking all the wrong questions. During her next call, a retail dealer requested immediate delivery on an inexpensive item that Jan's company distributed. Jan checked her inventory report and learned that the item was not in stock. She relayed this to the customer's dissatisfaction and explained that currently the demand for that item exceeded the supply. The retail dealer complained that they needed that item to complete a $10,000 purchase order and their customer would not accept delivery on partial orders.

Jan felt the pressure to satisfy this dealer. She called the warehouse manager, and there was no answer. She paged the warehouse manager, but no one responded. Then Jan's boss signaled that she had another important call coming in, so Jan asked the retail dealer to hold for a moment and picked up the other call. This call was going to take time, they wanted to reconcile last month's billing statement. She asked the second caller to hold for a

Figure 2

The Service Placebo

Warehouse

Warehouse Manager

Don't worry! Our warehouse manager is shipping your order as we speak.

moment so she could get back to the retail dealer, who by now demanded to know when the item would be available. Jan, taking a chance, told the retail dealer that stock on that item would be in next week. She hoped that would be enough time for the warehouse manager to restock the item. The retail dealer made business plans, based on Jan's promise, which included the $10,000 in revenue they expected from their customer's order.

The retail dealer called their customer to relay the fulfillment date. Their customer made decisions and plans based on having this merchandise in by the fulfillment date. The day after the fulfillment date, the merchandise still had not arrived and the retail dealer called Jan to complain. This time, Jan got through to the warehouse manager and learned that the item was manufactured by a vendor whose plant was closed for vacation. Production would not resume for another two weeks. Jan relayed this information to the retail dealer and apologized for the inconvenience. The retail dealer was furious. He had to change his business plan and he looked bad to his customer. So down the line, apologies and explanations were made, plans had to be changed and business relationships grew tense because Jan took a chance and hoped for the best. Jan learned an important lesson, it is better to disappoint customers with the truth, than to satisfy them with a lie.

Jan thought she had no choice. At the time, Jan was too pressured to get the correct information so, she gave her customer a placebo. The short term disappointment the retail dealer would have experienced, in knowing the item would not be available for weeks, was insignificant compared to the long term damage done to Jan's company and her own credibility.

To ensure a future for yourself in customer service, you must be honest. Speaking the truth is difficult when you're telling a customer what they don't want to hear. Notice I used the word *difficult* - not *impossible*. How you say something is just as important as what you say. Using the correct words makes all the difference.

It's not what the CSR says, it's what the customer hears. For example, telling a customer who accidentally breaks something "You broke it" may be the truth, but to the listener it sounds like an accusation. To remove all blame, focus on the real issue, and set up a helpful environment. One might leave out the *who did it* and concentrate on *what happened,* with a statement like "It will not operate under those conditions, allow me to help."

Words will either satisfy or antagonize our customers depending on their use and context. Disappointing customers with the truth by using the incorrect words will make things worse than they need to be.

Despite the best effort to keep mistakes to a minimum, problems will arise in business. The severity of business problems may differ from an inquiry regarding a billing error to a crisis that results in emotional behavior but the fact is; if there weren't any problems you would not have a job. With this in mind we accept the reality that sometimes things go wrong and when they do, the

Figure 3

result is often an upset or irate customer. As professional communicators, one of our jobs is to change someone who thinks he is our adversary into our ally.

The following guidelines are preventive. They will enable you to avoid the obstacles that make bad situations even worse.

Using "I" Instead of "You"

Correct words will enable you to communicate more professionally. The following examples of incorrect and correct phrases help contain adverse situations. Using them will help you avoid obstacles by not creating them in the first place. Notice that many of the correct phrases substitute the word "I" for "YOU". Think of the word "YOU" as a finger pointing the customer in the chest each time it is repeated.

Incorrect: You're wrong! It doesn't work that way.
Correct: I may be wrong, but I think it works a
little differently.

Incorrect: If you want my help, you'll have to...
Correct: I want to help, but first I'll need to...

Incorrect: You didn't do it correctly...
Correct: I came up with a different result. Let's look at
this together.

Incorrect: What's your name? or Who are you?
Correct: Who am I speaking with?

Incorrect: What's wrong?
Correct: How may I help you?

Incorrect: What did you say? or What!
Correct: I'm sorry, I didn't hear your last sentence.

Incorrect: You called too late. You'll have to wait until tomorrow.

Correct: I'm sorry, he/she has gone for the day. May I have him/her call you tomorrow?

Incorrect: You broke it!

Correct: It will not operate under those conditions.

Incorrect: You're confusing me!

Correct: I'm confused.

Incorrect: You'll have to read the manual if you want to learn how to do that.

Correct: The manual has an entire chapter on that, it's on page number...

Incorrect: You missed the deadline.

Correct: I'm sorry but the deadline has expired.

Incorrect: You sure do have a problem, you want some help?.

Correct: Please allow me to resolve this situation.

Accentuating the Positive

Focus on the positive potential of a situation not the negative possibilities. Saying, "I would hate to see" or "isn't as bad" focuses on the down side of a situation. Direct the customer's attention towards the positive potential for an optimistic future. Substitute either event, or situation for the word problem. Saying the word *problem* repeatedly only reinforces the fact that something is wrong. Saying *event* or *situation* will minimize the problem and keep your customer's thinking objectively.

Incorrect: This isn't as bad as the last problem.
Correct: This is much better than the last situation.

Incorrect: I am sorry you had to wait so long.
Correct: Thank you for being so patient.

Incorrect: Your problem is that we ran out of stock on that item.
Correct: Due to the high demand for that item we are temporarily out of stock.

Incorrect: You sure do have a problem.
Correct: I can help you resolve this situation.

Incorrect: Your problem is really serious.
Correct: This situation is a little different.

Incorrect: I would hate to see you experience a problem like this again
Correct: I feel confident that this situation will not arise again.

Incorrect: I would hate to give you wrong directions.
Correct: I want to give you correct directions.

Incorrect: You keep having problems with my company's equipment.
Correct: It appears these events are similar.

Incorrect: Don't forget to....
Correct: Please remember to....

Incorrect: I can't give you his extension number.
Correct: Only he can give out his extension number.

Editorializing, Errors, Absolute Extremes and Imperatives

Don't editorialize! Saying things like, "Yeah, we've had lots of these problems this month," only weakens your company's image. It also gives irate customers more ammunition to use against you. Handle each customer individually so they will think their problem is a separate event.

Some errors occur because a mailing address or an account number is misunderstood. Since it is in everyone's best interest to keep errors to a minimum, be aware of the consonants that sound similar on the telephone. Consonants such as T and P, or S and F sound the same. You should always say, "T as in Tom" or "S as in Sam" to clarify information.

To effectively deal with each customer, CSRs need to avoid using absolute extremes such as, *every, all, always* and *never* because they are probably not true, and they are antagonistic. In addition, avoid using the word "guarantee", in respect to what the future may bring. This usually happens when a customer asks, "What if this happens again?". Your response should not be, "Don't worry I guarantee it won't." CSRs can not predict the future or guarantee that adverse situations will not occur. Predicting the future performance of people, places and things is outside of a CSRs control. If you want to make guarantees, then guarantee your integrity, honesty and fairness. In the long term, customers want to work with someone they can trust, not someone who will make promises they can't keep.

Incorrect: I guarantee it will never break down again.
Correct: I feel confident in my work and feel free to call
            me if I can be of service again.

Incorrect: He's never at his desk.
Correct: He stepped away from his desk, may I have him
            return your call.

Incorrect: You always call at the worst possible time.
Correct: I'm working with another customer, may I call
you back when I am finished?

Incorrect: All you ever do is complain.
Correct: How may I help you today.

Avoid using imperative phrases when speaking with customers. To a customer, an imperative phrase sounds rude and commanding. These phrases makes you appear domineering, inflexible and dictatorial. Avoid using the word *listen* when attempting to capture your customer's attention. In addition, refrain from telling customers what *they need* to do for you to help them. Instead tell them what *you need* so you can help them. Doing this will prevent you from offending customers and enable you to be perceived as helpful, not antagonistic.

Incorrect: Listen! You won't find a better widget anywhere.
Correct: I believe, if you give this widget a chance, you'll
agree it is the best.

Incorrect: Listen it's not broken, all widgets work that
way.
Correct: It appears the widget is working fine.
Let's explore where the difficulty lies.

Incorrect: Listen! You have to do it today.
Correct: I would really appreciate if it was done today.

Incorrect: Sure I'll send you one, but you need to give me
your name and address.
Correct: Sure I'll send one immediately, may I please have
your name and address.

Incorrect: You don't understand, listen this time.
Correct: I may not be speaking clear enough, allow me to explain again.

Incorrect: I told him to do it today.
Correct: I asked him to do it today.

The Big Rush Off

Let's say you're speaking with an upset customer. He is venting his anger about a problem and all the inconveniences and embarrassment it has caused him. The customer is halfway through describing a problem you already know how to fix and you are in a rush, so you interrupt and tell him what he was going to tell you. This seems rational to you because you already know what the problem is and how to resolve it.

Realizing he is being rushed, the customer gets upset. Now you must spend another few minutes calming the customer down as he complains how he dislikes being interrupted. So in an attempt to expedite a resolution, you waste time, hurt the customer's feelings, and rudely represent your company's service.

CSRs should let the customer decide when to close a call. Rushing customers off the phone will undo all the right things you have done up to then. Some customers are not as busy as you are and they have a little time for polite conversation before closing a call. Allow them this courtesy as it will only take a few seconds and will leave them with a positive image of you and your company when the conversation ends.

Customers have thought about what they are going to say before they call you and until they get it off their chests, they will not feel serviced. To customers, this is a unique event. They don't realize that you hear their problem described different ways by many customers during the day. Always permit each customer to finish their description. Interrupting a customer only compounds a problem and any time you think you might save by doing this will actually drag out the resolution rather than expedite it.

While customers describe their problem, listen actively and acknowledge your presence by saying, "I understand" or "I see." Also, force yourself to breathe slowly and remain calm so when you do speak you will not sound rushed and anxious to get them off the phone in a hurry.

When you have other calls waiting for you, tell the truth. Be honest and say that you would like to acknowledge another call for a moment. When you pick up the other line, tell the waiting customer that you are presently working with another customer and when finished, you will return his call.

Answer each call as though it was your first call of the day, even though you may have been chewed-out on your last call. Allowing your frustration to become apparent to the next caller will not gain you any solace. Think of how you would feel, if after dialing customer service your call was answered by a desperate sounding CSR!

Verbose Customers

You may, from time to time, receive a call from a verbose customer who prolongs a phone call beyond the usual length of time. These customers need to be handled properly without being rushed off the phone. There are many reasons why some customers prolong a telephone call. Here are a few: (1) He called to ask a question and did not receive a satisfactory answer. So, the customer keeps a CSR tied up until the customer receives what he perceives to be a satisfactory answer. (2) A CSR may have provided a satisfactory answer and the customer is too embarrassed to admit that he doesn't understand what to do. So, the customer prolongs the call hoping to pick up some new information that will shed light on his dilemma. (3) The customer might just be lonely and have no one else to talk to or he may not know what it is he really wants. When handling a verbose customer, use the correct

---

words and guidelines discussed earlier in this chapter. In most cases, simply saying "I know you're busy, so I'll let you go." is an effective way to close a verbose customer.

Other situations might need more thought. There are a few things you can do. The first thing to do is ask the customer, "Have I answered all your questions today?" Asking this will force the customer to think about their original purpose for calling you. Your question may prompt another inquiry, but at least you are moving forward towards an ultimate goal - satisfying your customer and eventually closing the call. Second, use a soft compliment (people always listen when you compliment them). Try saying something like - "You seem to have picked up these new concepts rather quickly, I'm sure once you receive these materials, you'll be all set. However, if you need additional help after you receive these materials, feel free to call again." Doing this builds a customer's self esteem and gives them confidence to continue on their own.

Third, if the call is going nowhere, offer to get back to the customer on the following day. I have found that doing this gives the customer an opportunity to collect his thoughts so he can better explain his needs and concerns. Some customers have a difficult time thinking on the spot and they may need more time to determine what it is they really want. Since your customers are not professional communicators, you must use your expert communication skills to satisfy their needs.

Using Restraint not Retaliation

Eric is a customer service representative for the Quality Cup Company. During the last week, things weren't going so well. Eric had received two major complaints from ACE Distributors, one of his biggest accounts, about shipment shortages. Eric managed to contain both these events by using good communication skills and by responding fast.

Then, Randy from ACE Distributors called Eric about another mistake. Randy complained about receiving wrong merchandise. This time, he received 12oz cups instead of 15oz cups. Randy, speaking in an arrogant and condescending tone, reminded Eric about the last two problems, then Randy said, "You guys screwed up again!" Randy demanded an immediate replacement for the wrong merchandise. Eric, acting expeditiously, apologized and offered to ship 15oz cups overnight at no cost. Instead of being thankful, Randy was arrogant and he closed the call by saying, "Can't you guys do anything right."

Later that day, while reviewing ACE's original order for the 15oz cups, Eric noticed that ACE's purchase order specified 12oz cups. This meant that ACE made a mistake, not Quality Cup. Eric gloated! This was his chance to get back at Randy.

Eric called Randy and told him about the original purchase order and that ACE owed Quality Cup a reimbursement of $22.17 for the overnight freight charges. Randy refused to pay it and reminded Eric about the two previous problems that cost ACE lost time and revenue. Eric wouldn't budge. He wanted revenge! They got into a screaming match over the $22.17 and eventually Randy threatened to have ACE's president deal with Eric's company.

The next day, the president of ACE called Quality Cup and requested a return authorization for all their Quality Cup merchandise. He simply said that he had found another supplier.

Randy hurt Eric's feelings by exclaiming "Can't you guys do anything right?" Randy was wrong. Far worse, was Eric's retaliation with similar behavior. Eric, as the customer service professional, should have practiced restaint.

Customers become irate for a reason. Our role is to fix what is causing the customer to behave this way. Fix the problem, not the behavior. By depersonalizing abrasive remarks and focusing on what is driving the customer to behave this way, we stand a better chance of doing what's best.

Using restraint, rather than retaliation will enable you to contain situations, resolve the real problems and preserve long term business relationships.We sometimes must neglect our own concerns in order to satisfy the concerns of our customers. Your concerns might include only handling calls from customers that are respectful so you may work in a stress-free working environment. The customer's only concern is satisfaction, which sometimes means getting aggressive to get what he wants. Obviously, your concern is not the same as your customers'. When you are working with rude or abusive customers, use restraint.

An angry customer's behavior is unpleasant, but if you feel the temptation to retaliate, don't. This will only escalate a situation, not contain it. Behave like a professional. Speaking in a calm consistent tone will help the customer conform to your behavior.

If you retaliate, one of two things will happen. The customer may counterattack more severely and the situation will escalate out of control. Or, you will damage the customer's self-esteem and force him to submit for the short term, but lose the customer for the long term when next time he purchases some place else. As in everything else in life, the golden rule applies to customer service; treat others the way you wish to be treated.

Opening and Closing a Phone Call

The first few moments of a customer service call are special. It is during these few seconds that a customer creates a mental image of the CSR he is speaking with. Much of what a customer perceives and inevitably believes about you and your company will be set during the first few seconds. Will the customer think he's speaking with an alert, articulate professional or an apathetic, disinterested person? You have an opportunity to set the tone for a telephone call's outcome with a good start.

Our voice says much about our health, our attitude, our education and our assertiveness. With this in mind, open every call in a clear fresh tone and make sure you smile. If you don't feel like smiling, then try to fake it. An insincere smile is better than a sincere frown.

Figure 3A

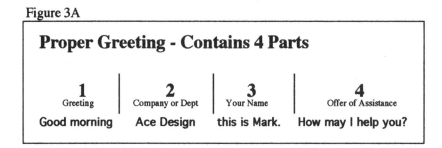

A proper telephone greeting should include the following four components: (1) Greeting, (2) Company or department, (3) Your name and an (4) An offer of assistance (see figure 3A). When a caller hears an opening that includes these four components, the telephone call is off to a good start.

Articulate your consonants and vowels appropriately. As a rule, consonants help to start and stop your vowels. Keep your consonants crisp and precise. Slurred speech arises when people allow their vowels to bleed into one another.

If you're prone to say "Uhm" often it's usually because you're not sure what to say next. People use "Uhm" as a place holder or filler word while they're thinking about their next word or phrase. I suggest slowing down your rate of speech so your brain can work faster than your mouth and provide you with a steady stream of words. In addition, avoid using common colloquialisms that might diminish your image. For example, replace "Yep," with "Yes," or "Certainly." Replace "Nope" with "No" or "Not at this time."

Figure 3B

Proper Closing - Ask a Question for Closure

Is there anything else I can help you with?

or

Have I answered all your questions today?

A proper closing should include a question that resolves the call and creates a mutual understanding of closure (see figure 3B). As you're getting ready to close a call, ask: "Is there anything else I can help you with?" or "Have I answered all your questions today?" These questions help to resolve the call on a positive note. When a customer hangs up their telephone receiver, at the conclusion of a phone call with you, they ought to feel special.

Chapter 1 — Key Points

- Avoid obstacles by not creating them in the first place.
- Honesty is essential in Customer Service.
- Placebo service doesn't work.
- It's better to disappoint a customer with the truth than to satisfy a customer with a lie.
- Use correct words that satisfy customers, not antagonize them.
- Refrain from editorializing, using absolute extremes and imperatives.
- Don't rush customers off the phone. Allow them to decide when the call is finished.
- Remind verbose customers of their initial purpose for calling your company.
- Use restraint, not retaliation when working with difficult customers.
- Avoid using common colloquialisms which might diminish your image.

Chapter 2:
Solving Problems

Looking rather exasperated, Jim Cook entered the electrical repair shop and dumped his VCR on the counter. When the repair technician asked what was wrong, Jim yelled, "IT'S BROKEN!" Speaking in a soft tone, the technician told Jim how difficult it was when appliances failed to meet our needs. Jim nodded with approval.

The technician complimented Jim for being smart enough to bring the VCR directly to the shop rather than the appliance store where it was purchased. The appliance store would have merely sent it to his shop anyway, and tacked on a handling charge. Jim liked being called smart. It made him feel good. Although Jim didn't know the technician very well, Jim thought he was an O.K. guy.

Feeling as though things were under control, the technician asked Jim a qualifying question, "What exactly is wrong with the VCR?" Jim answered, "The rewind button doesn't work." As he filled out the repair tag, the technician agreed to diagnose the VCR the following day and call Jim with a repair estimate.

The technician in this story used the Contain, Qualify, Correct method of problem solving. The technician, immediately realizing things were not under control, knew the situation would need to be contained before he could qualify the problem. Once the situation was contained and the customer was calm, the technician asked a qualifying question. With a specific problem description, the technician could now do his job.

Contain, Qualify, Correct is a problem solving method that enables CSRs to get things done right the first time. It increases productivity and customer satisfaction. Contain means you are getting the situation under control. By keeping the situation under control you will be able to do your job. As an expert in your field, you need accurate answers to your qualifying questions. Imagine trying to get accurate answers from an emotionally upset customer. These customers must first be calmed down.

Emotionally upset customers will not give you accurate answers. Instead, they'll use absolute extremes such as - "It *never* works" or "It's *always* broken."

In a survey of 200 of the nation's biggest companies, it was found that the skill most lacking in American workers isn't reading or computation, it's problem solving. This includes teamwork, interpersonal skills, oral communication and listening. Training in problem solving was offered by more companies than any other basic course.

The flowchart in Figure 4 illustrates how to contain, qualify and correct a problem. If you do not contain a situation you run the risk of misdiagnosing problems. Not solving problems right the first time means you may have to undo or redo work, and that may be costly. If this happens customers will blame you for the mishap, further complicating the situation, even though it was their behavior that confused you in the first place.

When situations are under control, you may qualify them, by asking questions to determine what is wrong. Asking questions allows you to control the conversation. Once you contain and qualify a situation, then you may begin to correct it. You must

Figure 4

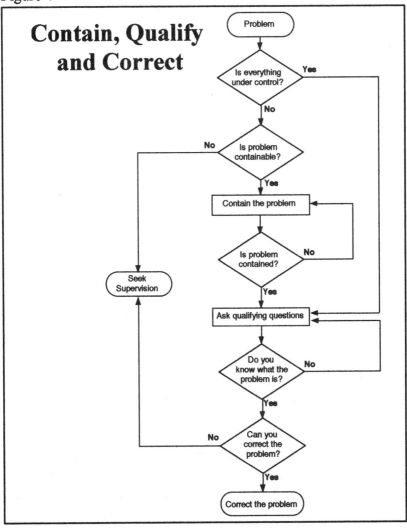

Contain, Qualify and Correct

Problem

Is everything under control? — Yes

No

Is problem containable? — No

Yes

Contain the problem

Is problem contained? — No

Yes

Seek Supervision

Ask qualifying questions

Do you know what the problem is? — No

Yes

Can you correct the problem? — No

Yes

Correct the problem

contain before you qualify and you must qualify before you correct. This method allows you to control events, rather than events controlling you.

During the contain process and the qualify process there is an option to seek supervision. This option is reserved for the exceptions that may arise. In addition, legal threats or extremely difficult calls should be passed up to your supervisor. Your supervisor has a broader scope of information to work from and can negotiate a mutually beneficial solution.

You can't get to the core of the problem until you defuse the customer's behavior. When a customer is irate, his metabolism is geared for battle. In this condition, customers may behave unpredictably. Different customers react to different situations in a variety of ways. What is a mild inconvenience for one person, may be a major catastrophe for another. This unpredictable behavior makes customer service extremely challenging.

Response Time

Response time is an important factor when solving problems. The longer you wait, the more difficult problems are to resolve (see Figure 5). If the technician in the previous story is three or four days late with the VCR repair, the customer will certainly become frustrated and possibly even irate. The time between the customer's report of a problem or a due date and the problem's resolution is directly proportional to its severity. Customers are more satisfied and problems are easier to solve when you respond immediately or resolve their problem on or ahead of schedule.

It's best to phone customers before they phone you. Doing this makes it easier for you to control the conversation since you initiated the call. You get to tell the customer what you are going to do, how you are going to do it and how long it will take. It keeps you on the offensive and the customer on the defensive. The opposite occurs when you do nothing. The customer is left brooding, until he finally goes ballistic and calls you.

When it comes to personal intervention, you must respond quickly instead of procrastinating until tomorrow or the day after. Never become complacent about getting personally involved in solving problems. Complacency occurs when after doing something well, due to much preparation, you see how much less you can do the next time and still achieve the same results. Don't get caught in the complacency trap. Call customers immediately and don't fool yourself into believing that today's customer will wait a little longer than yesterday's customer did. Yesterday's customer was satisfied with your quick response and personal attention. Why should today's customer be satisfied with anything less.

A recent study reports that responding to a complaint yields a 95% chance of customer retention. Not responding drops the retention rate to only 37%. To keep customers coming back, it's best to establish a system that ensures a quick and effective response.

Responding fast will satisfy customers and help you avert phone calls from irate customers. This strategy is preventive. It will help you to avoid obstacles by not creating them in the first place.

Figure 5

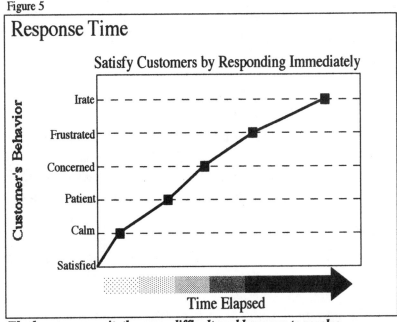

The longer you wait, the more difficult problems are to resolve.

Problem Solving Attitude

Greg is a customer service representative for a high tech manufacturing company. His company makes a specialized digital weight scale used by grocery stores. Greg noticed a trend in his calls as numerous grocers called in about a calibration problem. This problem resulted in a "CALIBRATION FAILURE" error message. "This is more than just a coincidence," Greg thought as he handled another complaint about this calibration problem. Greg reported these events to the engineering department. After reviewing several of the defective scales, the engineers found a connector problem on one of the scale's circuit boards. They designed an improved connector that would ensure greater continuity.

The company decided to do a factory recall to update all the scales out in the field. This made Greg angry. Now he had to contact all of his customers and inconvenience them with the bad news about a factory recall. "Why me?" Greg agonized. "If only the people in engineering had done their job correctly, the first time, I wouldn't have to go through this mess!" Greg's frustration was easily noticeable in his tone of voice and overall telephone manner. Many of his customers appreciated the pro-active phone message about the factory recall, but they didn't like Greg's poor attitude. As a result, many customers demanded that Greg's company simply buy back their digital scale so the customer could buy one from Greg's competition.

Greg was doing fine as a customer service rep until he started contacting customers about the factory recall.. He *contained* the problem well, then he *qualified* the trend by reporting it to the appropriate engineers. But, his attitude prevented him from *correcting* this problem effectively.

In most cases, customer service problems originate someplace else. It's true. For example, look at the different departments within a high technology company. These include a warehouse & shipping departments, the purchasing & manufacturing departments and the design and engineering departments. Each department has the potential to create or introduce problems that will result in customer complaints (see Figure 5A). The complaints might be something simple like a missing owner's manual or something more complex such as a design flaw. Either way, the engineer or shipping clerk who created the problem won't get to handle the customer's complaint. The customer service department will.

You might feel the urge to get frustrated about always having to clean up someone else's mess. Don't! That type of thinking is

Figure 5A

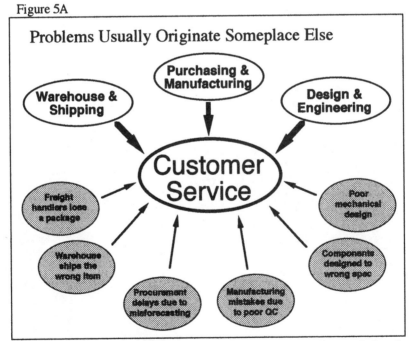

All forces, whether inside or outside a company result in work for the customer service department. Getting angry with other departments or customers will only cause relationships to grow tense and make for an adverse working environment.

inappropriate. It is very important that you fully grasped the nature of your job, which is to communicate effectively with customers and follow through to ensure satisfaction. You are a professional communicator. Mastering the art of communication is a very delicate exercise. It takes much objective thinking, a mastery of verbal and non-verbal clues and some gut instincts about what might be driving a customer to behave the way he is.

Using a problem solving method like Contain, Qualify and Correct along with an immediate response and a healthy problem solving attitude will enable you to resolve major catastrophes and minor inconveniences with consistent results. The next three chapters focus on how Contain, Qualify and Correct works.

Chapter 2 — Key Points

- Make sure things are under control before you attempt to resolve a problem.
- Once things are stable, ask qualifying questions to determine what is wrong.
- When a situation is contained and qualified, then you may correct it.
- Respond immediately. The longer you wait, the more difficult problems are to resolve.
- Be diligent about problem resolution. Don't get caught in the complacency trap.
- Maintain a good problem solving attitude regardless of where a problem originated.
- Fix the problem, not the blame.

Chapter 3:
The Contain Process

Contain means to hold back or restrain within fixed limits. When containing an event with an irate customer, your job is to stabilize the situation, so it doesn't get worse. When confronted by an angry or irate customer, your first reaction is to evaluate whether things are under control (see Figure 6).

Some customers will start by asking you to spell your first and last name. You should not feel defensive or be offended. Customers do this for two reasons, to hold you accountable and to intimidate. Don't allow them to taste their first blood!

Get There First With the Most

Customers looking to intimidate want to fight. They're being offensive to keep you on the defensive. A famous military tactician who fought for the Confederacy during the Civil War said the simple keys to victory are, "get there first with the most" and "always do what the enemy least expects." This strategy applies here (even though customers are not our enemy). The last thing a

Figure 6

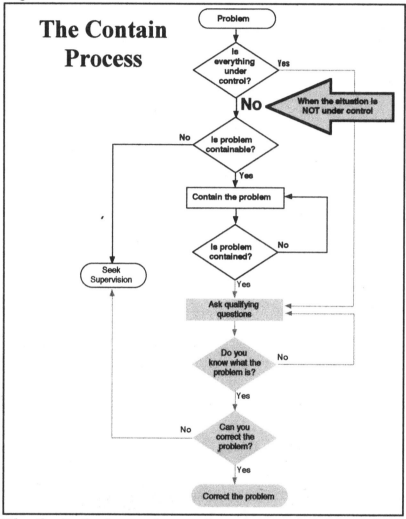

When the situation is not under control, you must contain before you begin asking qualifying questions.

customer expects is for you to spell your name for them, it's disarming. I have found it beneficial to spell my name for irate customers, before I am asked to do so. I even tell the customer to write my name down while I am spelling it for them.

Not all customers will ask you to spell your name, but do it each time. To those customers who weren't going to ask, it is a polite introduction. To those customers who were, it disarms their offensive action and interrupts their win/lose strategy. This strategy defuses the customer's offensive action. The event that a customer could chalk up as a victory, now becomes your personal introduction. *CONTAIN THE SITUATION*

Being genuinely concerned from the outset is critical. I usually start by asking, "Please, tell me what happened" or "Will you give me an opportunity to help you?" After this opening statement, listen to the customer's response. Customers that gripe are venting frustration. This is good. It means they trust you enough to tell you what happened and to give you an opportunity to resolve their problems.

Venting their emotions will usually take about a minute or two. When customers vent their feelings, you should listen actively and acknowledge your presence by saying, "I understand" or "Yes, I see". Refrain from saying "OK" whenever possible. Saying "OK" to acknowledge your presence is ambiguous to a customers who feels that things are not "OK."

When telecommunicating, unlike face to face encounters, it is possible to write while your customer is talking without appearing rude. Customer service representatives should always keep a pad on their desks for taking notes. So, put everything else aside and take notes. Write down pertinent facts like dates, names, etc.

Never interrupt irate customers when they are venting. Interrupting may send the message that what you have to say is more important than what they have to say. Irate customers need to talk about their problem so they can get it out of their system. Talking about their problem is like climbing a hill and when they get to the top, they'll feel better. Their story must be told in a contiguous sequence of events. When you interrupt an irate customer, they have start from the bottom of the hill again until they eventually peak. When the customer is finished venting, they will pause to take a breath. When they do, empathize with them.

I have a repertoire of memorized phrases that I use when empathizing. This is important because if the customer is really angry I am usually a little shaken up. In this state, it is difficult to think clearly and determine exactly how to appease the customer. These memorized phrases give me time to recover my composure and work out a game plan for satisfying the customer. I usually say, "I understand how frustrating it is when that happens. If that ever happened to me I wouldn't like it either." Empathizing with irate customers reinforces your similarities, not your differences. You might notice a change in the customer's attitude as they realize you're an ally, not an adversary. Now they may be receptive to permit you to serve them.

Using the information you documented while the customer was venting, verify the facts by reading them back to the customer. This will help keep the situation contained as the customer realizes you cared enough to pay attention and to understand the nature of the problem. Now you must control the conversation.

One of the best ways I know to control a conversation is to pay the other person a compliment. This will keep you talking and them listening. No one ever interrupts you when you are telling them how smart they are. Michael LeBoeuf, the author of the book, How to Win Customers and Keep Them for Life, wrote "Two groups of people always fall for flattery, men and women." Mr. LeBoeuf also writes, "A gossip is one who talks to you about other people. A bore is one who talks to you about himself. And a brilliant conversationalist is one who talks to you about yourself."[1]

I usually use soft compliments: "You did the smart thing by calling me because I am someone who can help you. I am glad you called because now we will be able to get things resolved"; or "You have a good sense of how to get things done. You also did a very good job getting your ideas across to me." Doing this reinforces your control of the conversation by establishing yourself as the expert. An expert, who can tell the difference between a good and bad presentation.

[1] Michael LeBoeuf, How to Win Customers and Keep them for Life, The Berkley Publishing Company© Page 43 and   Page 48

When you get stressed, your voice becomes raspy (unfortunately one of the many effects of stress). This change could send an irate customer the message that you are on the run or ready to cave in. Remember, the irate customer called to fight, and he wants to win. Keep your rate of speech slow and speak in a low pitch. Speaking slow will enable your brain to work faster than your mouth, so you might have the time to choose your words carefully. Speaking in a low pitch gives your voice the sound of authority.

Handling Vulgar Language

When containing situations with irate customers, bad manners are excusable, vulgarity is not. Customers who use excessive or vulgar language usually do so because of their emotional state. You should tell customers who use vulgar language, "I realize you are upset and I want to help you. But I am not in the habit of being spoken to in that fashion, nor do I speak with people who use that kind of language." This polite statement will keep the customer's self-esteem intact and allow the two of you to continue.

If the customer continues to be vulgar then it's time to disregard his self-esteem and make him feel silly. You should ask the customer to repeat the last sentence that contained the vulgar word. You should say, "Could you please repeat that last sentence, I am taking notes." The customer will think rationally about what they just said and will probably repeat the sentence minus the vulgar word. From that point on the customer should refrain from being vulgar. If in extreme cases, the customer insists they have the right to be vulgar then ask them if they would like to speak to your supervisor.

Customers that respond by being persistently difficult, making unreasonable requests and/or legal threats, may need to be passed on to your supervisor. As a supervisor, I believe customers are more afraid of me than I should be of them.

If an irate customer has had time to think about what they are going to say, they have also realized they may not get what they want. Before a customer calls, they have thought to themselves, "I hope somebody at the company helps me. What if the company can't solve my problem? What if no one at the company cares?" These fears go through the mind of all customers prior to their phone call. CSRs should do what the customer least expects; relieve the customer's fears and satisfy the customer's needs.

Chapter 3 — Key Points

- Allow difficult customers to vent their feelings without interruption.
- Spell your name for customers, before you are asked to.
- Listen actively and acknowledge your presence.
- Keep a pad on your desk to write pertinent facts on.
- Use soft compliments to soften-up the customer and position them for listening to you.
- Speak and breathe slowly to offset your body's response to stress.

Chapter 4:
The Qualify Process

Asking the Right Questions

During the last 10 to 15 years, there has been a tremendous change in high technology products and the way those products are serviced. Years ago, most electronic products relied on electronic hardware. These products were not very sophisticated, there were fewer features and a user, almost always, knew when the product malfunctioned. Likewise a service technician had an easier time troubleshooting these products as problems were simple to diagnose and correct. So, your kitchen appliances, televisions, audio/visual equipment used to be fairly easy to repair when the product failed.

Today, however, much of our usual electronic equipment integrates an operating system software in addition to the usual hardware. These products also include a plethora of features with a sophisticated user interface. It's difficult for a user to know when one of these products malfunctions because the interface is so complicated. Based on my interviews with CSRs, when a customer calls with a problem, the product is usually not broken. Therefore, CSRs must be expert qualifiers to ascertain whether a product is

Figure 7

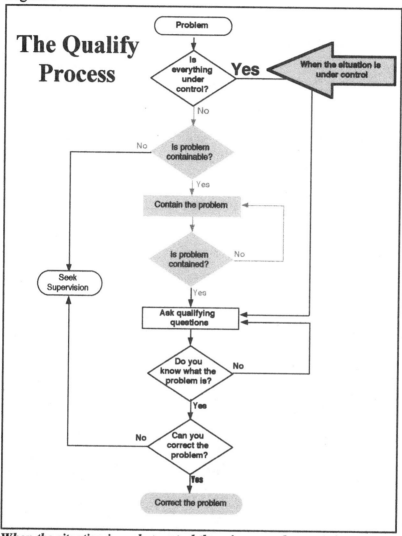

The Qualify Process

Problem

Is everything under control? — **Yes** — When the situation is under control

No

Is problem containable? — No

Yes

Contain the problem

Is problem contained? — No

Yes

Seek Supervision

Ask qualifying questions

Do you know what the problem is? — No

Yes

Can you correct the problem? — No

Yes

Correct the problem

When the situation is under control there is no need to contain, you may begin asking qualifying questions.

really broken of if the customer is experiencing operator error. Asking qualifying questions is the only way a CSR can get the information he needs to do his job.

During the qualify process, you'll need technical expertise in addition to your communicative skills. You may begin to qualify a situation only after you are sure things are under control. The qualification process will require asking the right questions. Good listening skills are important as some customers will not always respond with an honest answer as described in the next story.

Jack Campbell's microwave oven stopped working one day. He called the manufacturer and spoke with Murray, a technical support rep. Murray began qualifying the problem by asking Jack the preliminary questions like whether it was plugged in? Jack answered "Yes". Then Murray asked if the microwave's fuse was blown. Jack, noticing that the digital clock on the microwave was still working, assumed the fuse was OK and answered with a hesitant "Yes".

Noticing the hesitancy in Jack's voice, Murray surmised that Jack may not have actually qualified the fuse. So Murray asked Jack to remove the fuse and describe its condition. When Jack asked Murray where the fuse was, Murray realized that Jack had not visually confirmed the fuse's condition. Murray, not wanting to damage Jack's self-esteem, told him where the fuse was and upon visual inspection they learned that it was blown and in need of replacement.

Jack asked how the microwave's digital clock could work with a blown fuse. Murray replied that the digital clock ran off of an internal battery that lasted for years which enabled the digital clock to work even during a power failure.

This expert qualification saved Jack the time and inconvenience of a trip to a repair shop.

Some customers will say they understand something when they really don't. It's human nature, no one likes admitting that he doesn't know something, especially in a situation where the customer feels he needs a "win" for a win/lose result. Appearing incompetent may seem like a loss to the customer and make him even more confrontational. Since we are aiming for a win/win result, our job is to keep the customer's self-esteem intact and to avoid situations that might pressure customers into giving the wrong answer. Make the customer part of the solution rather than part of the problem.

Asking close ended questions provides the information you need to rule out *what* or *where* the problem isn't. As you narrow down the possibilities, continue asking questions that force the customer to answer either *yes, no,* or other *one word* responses. By asking open ended questions you may lose control of the conversation and the situation.

Imagine what would have happened if Murray, in the previous story, had not used the correct words. How would things have turned out if Murray said, "What's your problem?" when Jack first called in. Or, how do you think Jack would have felt if Murray said, "You're confusing me! What do you mean you don't know where the fuse is, you told me the fuse was OK." Or, what if Murray said, "You didn't really check the fuse, did you?"

The dialog between Murray and Jack could have become explosive if Murray hadn't used the correct words, as prescribed in chapter one, during the qualify process.

When asking qualifying questions, never provide the correct answer, and then ask the customer to confirm it. This is because sometimes, customers get unsettled during adverse situations and rational thinking may not always prevail. These customers may not admit they don't know the answer to a question, so they'll say they do when they really don't.

Ask questions that will allow customers to respond with one of the three following answers: (1) the correct answer, (2) the wrong answer, or (3) no answer in which case they'll say they don't know. This will enable you to guide the conversation and to determine whether the customer has a clear understanding of the situation.

Do not assume the customer knows as much as you do about your product or service. Therefore avoid asking questions that require expertise and ask questions that will give you the facts.

Some examples of correct and incorrect qualifying questions are:

Incorrect: Is the fuse blown?
Correct: What does the fuse look like?

Incorrect: When you press CTRL H on your computer
           keyboard, does the screen display a system error?
Correct: What happens when you press CTRL H?

Incorrect: Are you using the latest operating system?
Correct: What version operating system are you using?

Incorrect: Did you mail your payment on time?
Correct: On what date did you mail your payment?

The answers to the incorrect questions will provide a relative answer based on the customer's expertise. The answers to the correct questions will provide an absolute answer that reveals a customer's expertise. If you determine that your customer doesn't know something, use the correct words and inform him instead of blaming him. The situation will stay contained if you keep your customer's self-esteem intact. Embarrassing your customer might exacerbate the situation and make qualifying more difficult.

Separating the Customer's Problem from the Customer's Behavior

One morning, Ron Taylor, received a call from an irate customer. Ron was under verbal attack and his survival instincts were being provoked. After allowing the customer to vent, Ron asked the customer to hold for a moment to regain his composure. Ron went to his supervisor, Kaye Martin and said, "I have a customer on the phone, and he is bulldozing me!" Then, Ron explained that this customer has owned one of their products for a few years. "There is a problem and he won't let me help him. He said he wants his money back or else." Kaye asked Ron if he knew what the problem was? Ron said he didn't because the customer was too angry.

Kaye said, "A refund on a product purchased years ago isn't an option, and if the customer wants to speak with me about that, he can. However, it sounds like you have not qualified if there is a problem with the product yet. We should know that much before we do anything."

So, Ron took a deep breath, picked up the phone and reinforced his concern and willingness to help. He asked the customer some qualifying questions and within a minute Ron got results. Now using his technical expertise rather than his survival instincts, Ron had qualified that the customer was exceeding the limits of the product's capability. So Ron suggested a work-around which would allow the customer to continue using the product without a problem. The customer, appreciating Ron's assistance, apologized for his behavior and closed the call amicably.

The customer assumed the product was broken and he wasn't going to stand for it. So before the facts were in, he was unreasonable, demanding and irate. Ron's initial frustration was a result of reacting to and focusing on the customer's behavior.

Focusing on a customer's behavior may have forced Ron to lose restraint and to retaliate which would only have worsened an already bad situation. Asking the customer to hold for a moment allowed Ron to regain his composure and to seek some objective counseling from his supervisor. Ron was able to contain, qualify and correct this situation because he changed his focus from the perceived problem (the customer's behavior) to the real problem (the product). Separating the customer's behavior from the customer's problem is essential during adverse situations. Bad behavior is the result of a problem. Fix the real problem and your customer's behavior will improve.

Chapter 4 - Key Points

- Ask close ended questions that enable you to control the conversation and qualify the problem.
- Allow a customer to answer with either the right answer, the wrong answer or no answer.
- Don't assume the customer is an expert because he says so.
- Use the strategies discussed in chapter one, so the customer will remain compliant.
- When you need help, seek supervision or some objective counseling.
- Separate the customer's behavior from the customer's problem.
- Fix the problem, not the customer's behavior.

Chapter 5:
The Correct Process

When my six year old washing machine broke, I called the manufacturer for help. I spoke with Judy Walker, a factory service representative who dispatched a local factory repair person to fix my washing machine. The next day, Judy called me to follow up on the repair. She wanted to know if the repair person arrived promptly and did the job in a timely fashion. Judy's follow up call impressed me so much that instead of telling friends that my washing machine broke, I told them how great the manufacturer's service was. This experience taught me an important lesson in customer service. I learned that though my appliance worked flawlessly hundreds of times, it wasn't until it broke that I considered discussing it with friends. Judy's call eclipsed a negative event with a positive one, and I became a walking advertisement for Judy's company. This is why good customer service is so important!

Was my experience a lucky break for the manufacturer? I doubt it! Their system was planned, designed and implemented right down to the follow up phone call. My problem was contained with an immediate response, then qualified by an expert technician and finally, corrected with a personal touch. Their sole purpose in handling my situation, the way they did, was to get me to respond the way I did; now I'm their walking advertisement.

Following up with a customer after solving a problem or performing a service is a wonderful way to add the finishing touch. It allows the last and most memorable interaction between you and the customer to end on a positive note.

Your In-House System and Your Customer

During a usual work day, you handle a variety of calls from customers. Most of these calls funnel though your in-house system with little effort (see Figure 8). However, sometimes exceptional situations arise which may require some decision making by you or your manager. Where ever there are policies, there are exceptions. These situations are difficult because they won't flow through the funnel of your in-house system. But since people's lives are often disrupted by one thing or another, it is inevitable that things will not always go according to plan. Someone who is on their way for an important deadline may experience car trouble and have to cancel their appointment. Children may get sick and make it impossible for their parents to go to work or to make an important meeting. These situations are real and they need to be addressed one at a time. Always gather the facts of the exceptional situation and make the right decision for the right reason. Weigh the long term implications of your decision. No one would argue that staying home with a sick child is a good excuse for missing a deadline. Sure policies must be adhered to but sometimes these things happen.

Figure 8

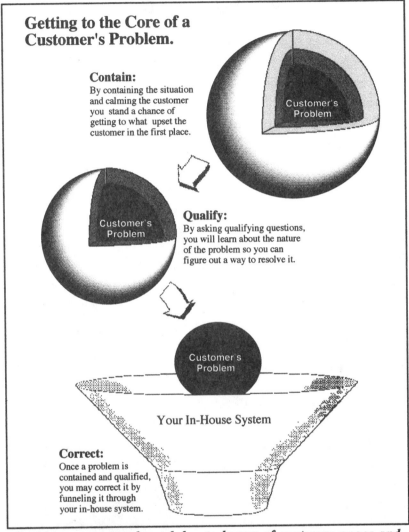

Getting to the Core of a Customer's Problem.

Contain:
By containing the situation and calming the customer you stand a chance of getting to what upset the customer in the first place.

Customer's Problem

Qualify:
By asking qualifying questions, you will learn about the nature of the problem so you can figure out a way to resolve it.

Customer's Problem

Customer's Problem

Your In-House System

Correct:
Once a problem is contained and qualified, you may correct it by funneling it through your in-house system.

Problems appear enlarged due to layers of customer anger and frustration. Getting these problems to flow through your in-house system will require a consistent problem solving method like Contain, Qualify and Correct.

Making right decisions 100% of the time may be impossible. I suggest that if you are not sure about what to do, seek objective counsel. When you are sure about a decision, go ahead and make it, providing you are working within the parameters of your job. As you gain decision making experience, you will speak with more confidence and authority. In customer service, we must not to be too rigid when a customer has an extenuating circumstance. All systems and policies need to have a little flexibility built into them.

Vilfredo Pareto was an Italian engineer who applied mathematical principles to social and economic phenomena in the early 1900's. His law of the "trivial many" and the "significant few" is referred to as the 80-20 rule. This rule applies to customer service because most customer complaints will come from a minority of your customer base. In-house systems handle the majority of situations, however, you must have a contingency plan to handle exceptions that may arise.

Calling an airline to book a flight is a routine task, for which an airline has an in-house system. By qualifying the passenger's destination, arrival and departure times the CSR handles and contains the situation. But, when a passenger calls from a gate to change his or her flight, this task gets more complicated for airline employees. They will first have to contain the situation by locating the passengers luggage. Then qualify that it has reached the amended airplane for completion of this revised scenario.

If something goes wrong and a passenger calls from her destination to complain that her luggage went to the original destination, the airline has a problem. This task will take some extra time and resources to resolve. In addition to locating and delivering the misplaced luggage, there is an irate passenger that will need placation.

In reality these things don't happen often, but when they do, there must be a contingency plan for exceptions. Whether unusual events like the airline example result from your company's mistakes or your customer's actions the result should be the same: solve the problem to the customer's satisfaction.

Once a situation has been contained and qualified it is time to correct. This usually requires a decision. Let each situation flow through the in-house system unless special circumstances make it an exception. Then make an informed, intelligent decision to satisfy your customer. Making good decisions requires good information. Always make sure you have all the facts regarding an event, so rational thinking will prevail. A person without facts is just a person with an opinion.

Once a problem is resolved, be sure to follow up with the customer to ensure satisfaction. Proper follow up enables you to (1) make sure everything is still OK, (2) it reminds the customer about your problem solving skills, and (3) it allows you to inform the customer about new products or services thereby maximizing your selling potential.

Chapter 5 — Key Points

- Follow up after resolving a problem or providing a service.
- Make exceptions when the circumstance warrants it.
- Making correct decisions is a learning process. Seek advice when you are not sure.
- Establish an in-house system for enhanced problem resolution.
- Make sure your policies and processes have a mechanism for handling exceptions.
- Get all the facts about an event prior to making a decision.

Chapter 6
Voice Mail Etiquette

The purpose of this chapter is to provide some basic voice mail skills that will enable you to stay productive when you are away from your desk. In addition, you will learn how to help others stay productive when you are unable to reach them while they're away from their desks. In this chapter, I will cover two aspects of voice mail etiquette: (1) that which involves the caller and how he ought to leave voice mail messages that facilitate work and momentum and (2) how a recipient should record voice mail greetings which invite callers to leave a voice mail message.

Voice mail is a tremendous productivity tool if used properly. My simple definition of voice mail is: A device which records and stores messages while you are away from your desk, along with a utility to retrieve the messages when you return. Let's stop here and review the previous definition. The key condition in this definition is *"while you are away from your desk."* This means that while you are at your desk, you ought to pick up the phone.

Ignoring incoming calls is not good business etiquette. Except for this chapter, the remainder of this book is meaningless if you don't pick up the phone. Rationalizing to oneself by insisting that it's OK to ignore incoming calls, because the call will be answered by your voice mail system, is just one more step down the slippery slope of inappropriate behavior. This type of behavior leads to a downward spiral of upset customers, lost sales and diminishing returns.

Since the inception of voice mail, callers have frowned upon having to leave a message with a machine. This perception further exasperates callers when their calls are not returned. This is unforgivable. Like any tool that is misused, voice mail could cause problems and waste valuable time and resources.

To reap the productivity benefits of voice mail, you must make callers feel comfortable about leaving you a voice mail message. Callers feel most comfortable when they believe, by past experience, that their call will be returned. This past experience affects their present behavior. People that are used to having their voice mail messages returned by you, will continue to leave you voice mail messages.

On the other hand, if you have a history of allowing messages to go unreturned, then callers will hesitate about leaving future messages. The absence of these voice mail messages will result in lost sales, less information, decreased productivity and fewer opportunities to build customer rapport. Unfortunately, many business people use voice mail as a way to hide from their customers or avoid work.

Garbage-In, Garbage-Out

Fran, needing to place an order, called Joe at Hex chemicals, Joe is not at his desk. His automated message says "Hi, this is Joe, I'm away from my desk, please leave a message." Hearing this cryptic and uninformative greeting, Fran instinctively presses "0" hoping to get to an operator who can page Joe. The operator's phone rings five times and then the call is forwarded to an automated attendant greeting with an

invitation to "Leave a message for a return call."
Impatiently, Fran hangs up the phone, checks her Rolodex
for another supplier and calls Ken at Best Chemicals.
Ken is Joe's competitor.

Upon calling, Fran learns that Ken is not at his desk.
Ken's voice mail greeting says, "Hi, this is Ken Truman
with Best Chemicals. Today is Tuesday, July 8th and I'll
be in the office most of the day. If you've reached this
message then I'm away from my desk, please select
extension number 214 and you'll be forwarded to Jill
Prescott, my associate, who will able to help you." Fran
dials the associate's extension number, gets through,
places the order and moves onto the next task at hand.
Meanwhile, Joe never realized he lost a sale due to his
poor voice mail etiquette.

The above story typifies a common interaction that occurs
thousands of times daily. Ken's voice mail greeting enabled Fran to
stay productive and finish her task. Joe's voice mail became an
obstacle. Listening to Joe's voice mail was an uninformative
experience - not so with Ken's voice mail message. Ken's message
started with the current date, which immediately implied that Ken
was actively involved with the day's business. By personalizing his
voice mail message every morning, Ken had learned that callers
are most likely to leave a voice mail message or try calling his
associate, thereby ensuring proper handling of customer's orders
and inquiries. Sharing the extension number of an associate gives
callers a fresh option to continue work in progress, rather than
yielding to an operator who may or may not know Ken's
whereabouts.

When Fran called Joe, she did not feel as if she was getting the
personal touch when, after seeking an operator, her call was
automatically forwarded to voice mail jail. In the world of voice
mail etiquette, this is incorrect. Transferring operator calls to a
voice mail box, without giving the callers a choice is wrong. When

Figure 8a

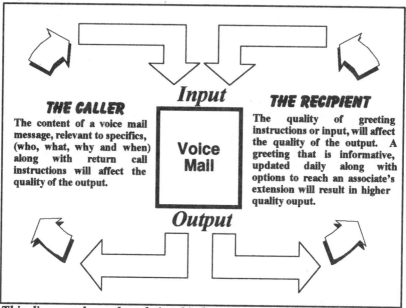

THE CALLER

The content of a voice mail message, relevant to specifics, (who, what, why and when) along with return call instructions will affect the quality of the output.

Input

Voice Mail

THE RECIPIENT

The quality of greeting instructions or input, will affect the quality of the output. A greeting that is informative, updated daily along with options to reach an associate's extension will result in higher quality ouput.

Output

This diagram shows the relationship between input and output voice mail parameters and how they might affect productivity.

callers have an expectation of reaching a human being, it's best to satisfy this expectation. Forwarding the call to an automated attendant is bound to frustrate and shed a poor light on you and your company.

I use the input/output paradigm to define the systemic qualities of voice mail. If you do not put the correct information into voice mail, you won't get the right information out of it. The garbage-in, garbage-out principle applies to voice mail, as it does to most other input/output devices.

Callers who get forwarded to voice mail are providing the necessary input for a voice mail system (see figure 8a). The content of this voice mail message, relevant to specifics, (who, what, why and when) along with return call instructions will affect the quality of the return call or the output. For example, if a caller leaves a cryptic message (garbage in), the recipient will not have the necessary information to do his job (garbage out). This results in the recipient making a return call so the facts can be surfaced.

These facts could have been relayed during the initial voice mail message, This would have saved time and increased productivity.

On the other hand, if the caller leaves a rambling and verbose message, without any meaningful content, the recipient has to suffer through the entire message and still not have all necessary information (garbage in). Again, the recipient will have to make a return call to get the facts. Either way, the recipient does NOT have enough information to do his job (garbage out).

The recipient has his own set of input and output parameters. The quality of his greeting instructions, or input, will affect the quality of his caller's messages (the output). For example, a greeting that is informative, updated daily along with options to reach an associate's extension will result in higher quality messages (the output) than a greeting without those attributes.

Do It Right the First Time

John is a logistics manager for a high tech computer manufacturer. On Tuesday morning, he was so busy that the time flew by. As he prepared to go to lunch, he knew his entire afternoon would be consumed with shipping allocations for a large far-east customer.

When John returned from lunch, he noticed that his voice mail indicator light was lit. So before John started his shipping allocations, he decided to check his voice mail. The voice mail was from a vendor. John listened, "Hi John, it's Laurie from Action Logistics. I need a favor. I hope you can help. Give me a call." John shook his head in disappointment. How was he supposed to know what Laurie wanted. Now, John had to call back just to find out what the favor was. Had Laurie been more informative, he would have simply answered her question with a return call. But now, he was obligated to call Laurie just to surface the facts.

John dialed Laurie's phone number and learned that

Figure 8b

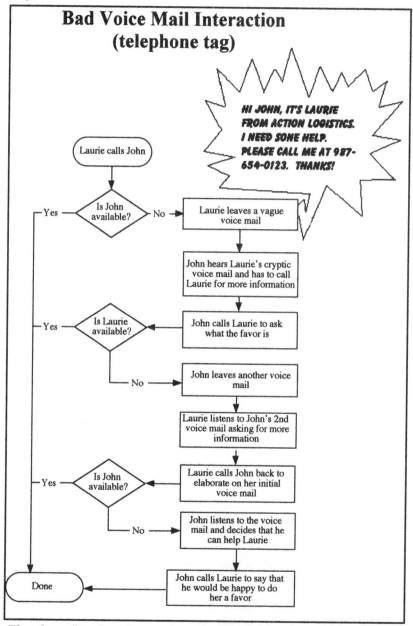

The above flowchart is complex and confusing because of all the extra steps. Incomplete and poorly executed voice mail messages result in a wasteful game of telephone tag.

Figure 8c

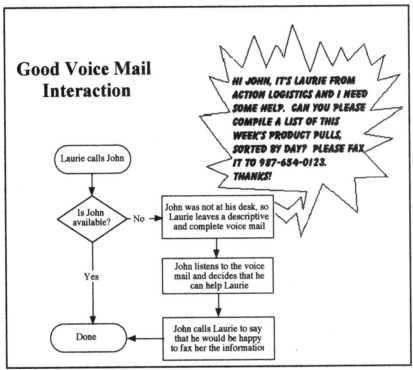

A complete and well executed voice mail message results in fewer steps which save time and increase productivity.

she wasn't at her desk. So, he left her a voice mail knowing that she would have to call him back thereby starting a frustrating and wasteful game of telephone tag.

Situations similar to the above occur often. If you consider the fact that each day, thousands of recipients like John, receive incomplete and poorly executed voice mail messages that require extra follow up. If the extra follow up takes an additional five to ten minutes and this time is multiplied by all the mishandled daily voice mail, the cumulative waste of time is astonishing. The flowchart on page 72 (figure 8b) represents the multiple steps required to satisfy a simple request Had Laurie asked her question in the initial voice mail message, as shown in figure 8c, then John could have simply replied with an answer. This is no exaggeration. Consider this the next time you record a voice mail message.

To make the most of your voice mail be sure to update your personal greeting regularly. A proper voice mail greeting should include the following five elements: (1) Your name, (2) the day of the week, (3) what touch tones a caller can press for immediate assitance, (4) when you'll be returning calls and (5) an option to reach the operator or receptionist.

Chapter 6 — Key Points

- While you are at your desk, pick up the phone. Don't ignore incoming phone calls simply because you have voice mail.
- When callers leave you a voice mail message, always return their calls. Allowing voice mail messages to go unreturned is inappropriate.
- Transferring operator calls to a voice mail box, without giving the caller a choice, is inappropriate.
- Ensure that your voice mail system doesn't allow callers to get stuck in a dead-end of voice mail jail.
- Refrain from leaving voice mail messages that are cryptic and uninformative.

Chapter 7:
Handling E-mail and Letters

Anyone in customer service, technical support and consumer relations has most likely had to start handling e-mail inquiries. The Internet explosion, during the last few years, has resulted in yet another way for customers to contact us. Handling e-mail is as delicate as speaking over the phone in that words are still weapons. In addition, most people write more abrasively than they speak and this escalates problems. For this reason, it's best to make sure you stay within the guidelines contained in the first chapter of this book.

E-mail offers a few benefits over telephone communication. For example, customer service departments that handle both telephone calls and e-mail, in most cases can defer the e-mail to a time of the day when the phones aren't so busy. In addition, if your company gets the same frequently asked questions (FAQ) every day, e-mail allows you to paste reply "nuggets" as a responseto a customer's e-mail inquiry. It's always best to personalize the opening of a reply nugget with a sentence that is relevant to the customer's particular situation. This will make the nuggets appear friendlier. Typical FAQ nuggets include stock answers to questions such as, "Where can I see one of

Figure 8A

When you write e-mail - Slow Down

```
Date: 01-Jan-1998
To: john@anywhere.com
From: coscia@worldnet.att.net
Subject: Slow Down

John,

Typing too fast might result in numerous
typographical errors.  Making too many
mistakes will diminish your image and
possibly create meanings that you didn't
intend.

Always write, review and revise each e-mail
carefully.

Best Regards,
Steve Coscia
Author
```

your company's products?" or "How much does it cost?" or "When will the product be available?"

E-mail vs Telephone

With e-mail, it's difficult to communicate as broad a range of information as in telephone interactions. In e-mail, the words come across OK, but you lose the verbal signals such as vocal inflections, vocal melody, pauses, tone and volume. These verbal signals help us interpret a meaning that lies beneath the words. With this in mind, remember that people don't take meaning from your words, people give meaning to your words.

Unlike telephone conversations, e-mail is often forwarded, copied or pasted over and over again. This means that your e-mail may come back to haunt you some day. Early on, I learned this lesson the hard way. It's best to be attentive and measure your words carefully. Never send e-mail when you're angry. First, cool down and think objectively about whether you really need to send it. Most e-mail applications will let you save a message for sending later. If you're not sure about whether to send a controversial e-mail - don't! Save the message and do something else for a while. When you return to the controversial e-mail, you'll be thinking more objectively.

E-mail Conventions

Each e-mail message should include a subject, a greeting, a body of text and a closing. The subject is a line at the opening of an e-mail message that allows you to tell the receiver what you're writing about. It's best to keep your e-mail subject short and descriptive. Along with using a descriptive subject, start the body of your e-mail message with a signal of what you're addressing or writing about. This helps the reader to orient themselves to your message.

Blank lines between paragraphs make an e-mail easier to read. Avoid using a tab to indent the beginning of a paragraph.

Figure 8B

Don't Use All Capital Letters

```
Date: 01-Jan-1998
To: john@anywhere.com
From: coscia@worldnet.att.net
Subject: Using Capital Letters

John,

NEVER DO THIS! IT LOOKS LIKE YOU'RE YELLING!

When writing e-mail, use ALL CAPITAL LETTERS
sparingly so your readers won't think you're
yelling at them.

Best Regards,
Steve Coscia
Author
```

Not all e-mail applications will work with tabs and doing so might yield unpredictable results.

Never write in all capital letters. In e-mail, doing this means you're shouting. Use all capital letters sparingly. Use the exclamation point "!" sparingly also. You can only generate so much emotion with the "!". Don't do this !!!!!!!!!!!!!!!!

Slow down when you write e-mail. Since e-mail is such an instant medium, it provokes people to write faster and sloppily. If your message is a mess, you will make your reader work harder than he needs to. Always review and revise an e-mail message before sending it. This becomes very important when you are writing about absolute times and dates. You might inadvertently make mistakes. Slow down and make sure your e-mail is accurate and well written.

Always sign off your e-mail with your name, title and company name. Don't use your company's name only. Keep e-mail as personal as you can.

E-mail Abbreviations

Among the many e-mail conventions, use of abbreviations is one of the most common. Using abbreviations saves keystrokes buy may add confusion if your reader is not familiar with their meaning. Listed in the following table are some of the more common abbreviations.

Figure 8C

Common E-Mail Abbreviations

Abbreviation	Definition
BTW	by the way
FWIW	for what it's worth
FYI	for your information
IMHO	in my humble opinion
RTFM	read the fine manual

Had I listed all the different e-mail abbreviations the list in Figure 8C would be much longer. In general, abbreviation usage is quite rampant on the Internet and some industries have created esoteric abbreviations for terms within their own markets. I suggest using only the most common abbreviations and avoid colloquial or esoteric ones. In customer service, it's best to not run the risk of confusing your recipients.

Angry E-Mail

I learned an important lesson about handling e-mail from angry customers not long after I went on-line. I received an e-mail from an angry customer. I replied with an offer to help and an explanation of my company's service policy and all I received was another e-mail with more anger and frustration. I replied again, with a verbose and elaborate e-mail (that took a significant time to write) that I thought would clear this matter up. Instead, the customer replied with more anger and frustration. I was getting nowhere. Finally, I e-mailed the customer and asked for her phone number and a "best time" that I could call her on my dime. She replied with her phone number and a best time. I called and resolved the situation.

When handling an e-mail from an irate customer, it's often best to reply with a message asking for the customer's phone number and the best time to call. Then, follow through with the phone call. In my experience, customers like my willingness to invest in a telephone call. My vocal attributes and verbal signals go way beyond what I can accomplish in writing.

Handling Complaint Letters

While e-mail is the preferred medium for written correspondence, some customers still send letters the old fashioned

Figure 9

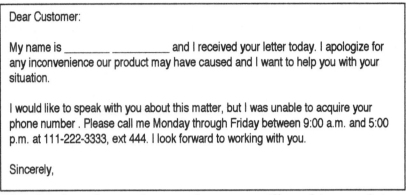

Dear Customer:

My name is _____ _____ and I received your letter today. I apologize for any inconvenience our product may have caused and I want to help you with your situation.

I would like to speak with you about this matter, but I was unable to acquire your phone number . Please call me Monday through Friday between 9:00 a.m. and 5:00 p.m. at 111-222-3333, ext 444. I look forward to working with you.

Sincerely,

Use the above letter when you know you will need to negotiate a settlement. If the customer merely needs an answer to a question, then include the answer in the letter..

way, via the post office. Most customers write letters to inform. They don't want a response. These customers merely want to make you aware of an unfortunate event or some other negative situation. Call them anyway! You'll impress these customers with your excellent follow up and concern.

Some of these letters may include scathing criticism of your company, demands for monetary compensation or threats to pursue legal and/or government agency assistance. In my experience, the customers who write complaint letters probably don't expect results. This is because customers always sound so shocked when I call them in response to their letters. My call eclipses the problem and any unreasonable demands made by the customer.

The rule for responding to letters is to call the customer immediately. If the customer does not include her phone number, call informati (dial area code + 555-1212 or go to the URL www.555-1212.com)) and if possible, get it. Prior to your phone call to the customer, take a few minutes to write down key points and phrases you want to cover during the phone call. Preparing for call-outs will prevent you from fishing for the correct words during your presentation. By investing a few minutes in preparation, your voice will sound more confident and authoritative. In addition, the

Figure 10

Dear Sir/Madam:

I am the owner of one of your widgets that has been plagued with many problems.

My experience with your company is limited to this one experience, but it has made me suspicious and extremely disappointed in your products. The only way to restore my faith in your companyis to replace my old widget with the newer, more expensive model your company recently released.

To keep me satisfied, extreme measures must be taken and an entirely new widget must be delivered to me. Please do your best to change my view of your company.

The Customer

Figure 11

Dear Sir/Madam:

I am writing this letter because I have had it with this widget, as well as the service and technical support.

Since day one the widget has malfunctioned. At the beginning I thought it was something that I overlooked, being new to the product and all, but over time the widget's problems have become worse and I realized it needed service.

After receiving my widget from repair, I began trying to finish the project I had started only to find out a few weeks later that the widget still did not work correctly. Once again I am being inconvenienced by your product. I expect that in the spirit of good business, as well as good faith, that you will compensate my business by doing one of the following:

- Repair my widget + $2000 (two thousand dollars)

- A new widget + 24 hour 1-800 Tech Support

- $5000 (five thousand dollars and we go our separate ways)

I need to hear from you within two weeks from your post mark date as to your intentions or else I will take further action.

The Customer

In both of the above examples, I phoned the customers immediately and arranged repairs for their products. Both customers were very satisfied and neither of them persued their unreasonable demands because they received prompt, effective service.

customer will realize your sense of purpose and mission which will boost you and your company's image.

Some customers make unreasonable demands in writing because they are frustrated. Some customers will even make legal threats. In my experience, the formula for handling these situations is; (1) to respond immediately, (2) to thank them for writing, (3) to compliment them on their writing style, (4) to focus on the real problem — not the customer's unreasonable demands and get the real problem resolved. Applying my formula appeases the customer's frustration, anger and unreasonable demands. Basically, customers want their problems solved. If you can do that, then you are satisfying their needs. I have found that most people are honest, they really want to be fair, and do not want to take advantage of your company. But, they also want to be treated fairly. If they feel they are not being treated fairly, they might respond by being unreasonable, just to get your attention.

While CSRs should always attempt to answer a complaint letter with a phone call, a customer will sometimes have an unlisted telephone number. If this situation occurs, then write them a letter asking them to contact you immediately. Do not attempt to negotiate or resolve a customer's problem with your letter. Use the letter to get them to call you. Then, using your expert communication skills, you will have a much better chance of negotiating a win/win result. Send a letter similar to the one in Figure 9. If the customer needs an absolute answer, such as an account balance, include that answer in the letter.

As I described earlier in this chapter, my customers are very surprised when I call in response to their letters. Their reactions are so positive that some customers apologize for complaining in the first place.

I have received letters with hostile demands for unreasonable monetary compensation or threats to pursue legal and/or government agency assistance. The outcome of almost every one of these events was extremely positive. The customer's situation, in almost every case, was resolved through our in-house system via usual channels. I believe customers exaggerate their demands in an effort to get someone's attention, not to be unreasonable and greedy.

The following examples of customer complaint letters in Figure 10 and Figure 11 are composites of actual letters. The only words edited in both letters is the name of the product, which I changed to widgets (no one knows what a widget is). In both situations, I called immediately and offered to repair the defective product at my expense, no monetary or material compensation, just good, prompt and personal service. Why do customers make unreasonable demands in writing and then when phoned, seem so easy to work with? They do it to get your attention. If you're attentive beyond their expectations by using my formula, fairness will prevail.

Complaint letters like those previously mentioned are not the only correspondence you will have with customers. You will, from time to time, receive wonderful "Thank You" letters full of praise and adoration of your expert customer service skills. I suggest saving these letters so you can review them from time to time, especially when you feel that things couldn't be worse. Keeping an objective attitude about customer service is the key to longevity in this business.

Handling Letters from the Better Business Bureau

In these litigious times, people have never been more aware of their rights. Consumers are no exception. Consumer advocacy agencies function as a mediator between a business and a customer.

If a customer has a gripe about a company, the customer can fill out a claim with the Better Business Bureau. All claims are screened for customers with unreasonable demands who are using the Better Business Bureau to help them get what they want.

If one of your customer ever files a claim with the Better

Figure 12

Dear Sir/Madam:

We recognize that there are two sides to a dispute and the complaint only shows the customer's version. The BBB needs your interpretation of the facts set forth in the complaint so that we may successfully mediate a mutually satisfactory solution.

Please send us a WRITTEN reply to the enclosed complaint within three weeks, addressing the issues raised by your customer. Please do not telephone us as this will merely slow down our mediation process. You may use the reverse side of this letter for your response.

BBB performance reports are based, in part, on companies' good faith reponses to valid customer complaints presented and processed by this Bureau. Your response to the enclosed complaint will be reflected in our Performance Report on your firm which is issued to all inquirers upon request.

The Bureau staff look forward to working with you in mediating a prompt solution to this complaint. If you do not hear from us after you have responded, it will indicate that your response is adequate and the complaint is closed. We thank you for your cooperation in replying, IN WRITING, within three weeks.

Sincerely,
Better Business Bureau

Business Bureau against your company, the Better Business Bureau will send you a letter similar to the one in Figure 12. A prompt response to this letter will ensure your company a positive rating with the Better Business Bureau. In your response, stick to the facts and don't editorialize. Express your willingness to work with the customer. Your company will not be judged or criticized for trying to work with your customer. The Better Business Bureau does not punish companies when their customers complain. Their function is to help customers and companies work together.

In an interview with a Director of Trade Practices/Mediation at the Better Business Bureau, I learned that about 90% of all complaints result from lack of good communication regarding a sales or refund policy.

Responding with the time allowed by the Better Business Bureau is important as after the deadline, a mediator will get involved to resolve the dispute. My advice is to respond immediately so the situation will not go that far. When you consider that one dissatisfied customer might tell hundreds of other potential customers that you didn't care enough to respond and were uncooperative, it's not worth it.

Keep a journal or note pad on your desk at all times to write down important information that might come in handy when researching facts for a response. I hope your customers never have the need to file a claim against your company. If they do, however, work within the parameters of the Better Business Bureau and try to win them back.

Chapter 7 — Key Points

- When writing e-mail review and revise your text before sending a message.
- Never write in all capital letters. Doing this means you're shouting and is considered rude.
- Use abbreviations sparingly to avoid confusing recipients.
- When handling complaint letters: (1) respond immediately, (2) thank them for writing, (3) compliment their writing style and (4) focus on the real problem.
- When customers don't include a phone number in their complaint letter, call 555-1212 and get it.
- If your customer ever contacts the Better Business Bureau, do the following: (1) respond immediately, (2) stick to the facts and (3) express your willingness to help.

Chapter 8:
Handling the Stress of Customer Service

Caroline Clark is a technical support representative for a computer software company. By 10:00 in the morning, Caroline had already handled about 15 calls. These calls ran the gamut from usual user questions to angry customers whose operational errors caused software problems. Caroline tactfully handles each customer with grace and respect.

By 11:00 a.m., Caroline had handled another 10 or 15 calls. The customers were more of the same, some technical, some emotional. By now Caroline is controlling her breathing and rate of speech to avoid sounding stressed out.

Her next caller at 11:05 a.m. is angry. This customer lost a file that was extremely valuable. When Caroline asked if he had a backup copy of the file, the customer answered no. So Caroline explained the benefits of doing regular backups so lost files could be restored. The customer knows Caroline is right but he doesn't make it easy for her. Before he hangs up he makes his displeasure known in detail about how files should never get lost in the first place. Caroline thanks the customer for calling, closes her eyes, takes a deep breath, composes herself and picks up the next incoming line.

Stress is a national epidemic in America. More employers and experts are realizing that stress, burnout and depression can affect productivity, absenteeism, turnover and overall health costs. The National Institute of Mental Health estimates that about 10 million Americans suffer from depression annually. The cost to the nation is more than $27 billion, with lost productivity of $17 billion just from absenteeism. A recent survey reveals that more than three-fourths of all workers say their jobs cause stress. Workman compensation awards for stress, almost unheard of a decade ago, now represent one-fourth of all claims. In addition, one-third of all patients seeking medical help have physical complaints as a result of stress.

Caroline manages the stress of her job by using the contain, qualify and correct method of problem solving and by thinking rationally. She focuses on the root cause of what makes some customers upset. These customers assume the only way to get satisfaction is to be loud and threatening. They feel no need to behave calmly when they perceive your company's product or service to be problematic. Caroline has learned that focusing on an angry customer's behavior is an obstacle. Their behavior is a result of the problem. Fix the problem, and you will automatically fix the behavior. For a more detailed description of stress and its effects on our metabolism, read my book TELE-Stress.

Irate customers may disrupt our well being and instigate the stress mechanism because, the last time you handled an irate customer, the experience was unpleasant. You may have agonized over the details in your mind and replayed the experience repeatedly. The situation was difficult, but it was over in no time. You dragged it out into a bad emotional experience.

Situations may be unpleasant, even dangerous, but it is how

Figure 13

Stress sneaks up on You

you perceive them that will determine whether they are stressful. Stress sneaks up on you during the course of your work day (see Figure 13) but you would have to stop living to avoid adversity. Adverse situations do not cause stress. If this was true, everyone would experience the same degree of stress. Since no two people have the same physical and emotional make-up, no two people will perceive adverse situations in the same way. What I perceive as a mild inconvenience, you may perceive as a major catastrophe.

Thinking rationally about stress and challenging our own thought process can help overcome the effects of stress in customer service. Dr. Samuel Klareich, author of Work Without Stress wrote this about stress; "Our society is very big on not thinking. We are very big on doing. In almost every major organization and corporation, the emphasis is on getting the job done. The emphasis is on problem solving, on completing tasks. Action is seen as the first and foremost activity. Thinking is secondary."[2] But, since irrational thinking produces stress, doesn't it make sense that rational thinking will manage it?

[2] Dr. Samuel H. Klareich, PH.D., Work Without Stress, Brunner/Mazel Inc. ©, Page 27

Let's look at the effect stress has on our bodies. Our general metabolism increases along with our heart rate, blood pressure, rate of breathing, muscular tension, even our voice changes. Most of these changes are good in the presence of danger. A certain amount of stress is positive and pleasurable. It inspires ideas, challenges and leads to productivity in the human race. In other words, stress enables you to meet deadlines, jump out of the way of a speeding car and handle life's many crises. Allow me to share two personal experiences that may help you understand our physical and emotional response to stress.

Physical and Emotional Stressors

I enjoy regular workouts and do a good portion of jogging in my own neighborhood. There is a five mile course I run some Saturday mornings. I enjoy this course because about halfway through there is a pond that I encircle. About 15 to 20 Canadian geese reside there and I usually run among them. One Saturday morning, while running this 5 mile course, I noticed a new house being built beside the pond. The frame had been erected on a new concrete foundation. I believe the geese were annoyed their turf had been intruded, and changed how they felt about people, because these geese seemed tense. Well, terror hadn't set in yet because my experience told me that this was not a dangerous situation. After all, I have been doing this run for about four years. Then, one of them began chasing me, and I got concerned, because he was moving faster than I was. Then the whole flock took off after me. I was terrorized! One of the geese flew up, hit me in the head and knocked me over. Although I was tired and out of breath due to my run, I got up and took off as though I were fresh and rested. I eventually made it back to the main road and resumed my usual run. This was a classic case of real physical harm provoking my survival instincts to react.

My second story is one that more of you may relate to. This

event occurred in the early stages of my career in customer service. One Monday morning I came in to work fresh and rested. I poured my first cup of coffee, you know, the one that brings that warm satisfaction. On the way back to my desk I heard my phone ring. I crash landed in my chair, put down my coffee cup and picked up the phone. On the other end was an irate customer. He had bought one of my company's products on Friday night, had a problem with it on Saturday morning and hadn't been able to use it all weekend. This customer had two days and nights to think about what he would say and do when he got me on the phone Monday morning. My physical metabolism went from calm to stressed in seconds. My stress mechanism was activated without any threat of physical harm.

The same energy that allowed me to get up and run away from the geese was pent up inside me during the dreaded Monday morning call and I had no physical outlet. I wanted to fight with that customer, not help him.

Retrospectively, my experience with the geese was something that could happen once a week, if I were brave enough to return to

Figure 14

Anxiety impedes your Performance

the pond. But, in customer service you could have your stress mechanism chronically activated 5 or 10 times a day. This could affect your well being and job performance. And this is inappropriate. Because, when your stress mechanisms are triggered many times a day, it is like riding on an emotional roller coaster. This can result in anxiety, irrational thinking, long term health problems and burn out (see Figure 14). So what do you do?

You can't perform well as a CSR if you can't handle stress. And there is no cure for stress. Physical exercise is wonderful for short term relief. It is a good outlet for pent up anger and frustration. Other short term relievers of stress are relaxation, meditation, walking, aerobics — there are many. These activities are wonderful for our muscular tone, cardiovascular system and mental discipline. However, they will not cure stress. But there are some rational techniques that, when applied, change your perception of dangerous or harmful situations.

You are sometimes your own worst enemy when it comes to thinking. You may think to yourself, "If that ever happens to me again, I won't be able to bear it." You convince yourself that some situations are unbearable, and give up. Situations are not unbearable, it is how you respond to those situations that makes them unbearable. If thinking is critical in producing stress, it is also critical in managing stress.

Think about what happens when you respond to a situation. You scan your memory bank to determine how you responded the last time a similar event occurred. This is how everyone learns about life. Our subconscious is programmed by experience to connect certain feelings with situations that seem similar. If an earlier situation was unpleasant, new situations that seem similar will evoke a similar feeling. This means that if you have bad memories of handling irate customers, future similar situations will result in fear that will invoke the stress mechanism. Adverse situations are manageable when you respond, not react. How you respond depends on your life experience. You must handle stress effectively when working with irate customers.

Stressors Within Your Company

Jeff Austin had to leave his desk for a moment and photocopy a technical sheet for a customer. In the corridor, he met Mary who reminded Jeff about a report deadline that was due tomorrow. Jeff remembered that he needed some research documentation before he could complete that report. "I might as well get it now", he thought. So instead of going to the copy machine he headed for the documentation department.

On the way, Jeff was paged on the PA system. He answered the page and was told to return to his department immediately for an urgent call. Wanting to get the documentation first, he asked to have the call switched to the phone he was at. He took the call, wrote notes on the back of the page to be photocopied and promised to call the customer in a few minutes with a response.

When he got to the documentation department, Jeff couldn't find what he needed. The clerk told Jeff that someone from engineering was using the documentation and would return it sometime this afternoon. Jeff said he would return later and remembering why he left his desk in the first place, went to photocopy that technical sheet. Unfortunately, the photocopy machine was being serviced and wouldn't be available for another few hours.

Jeff went back to his desk, without his photocopy or the documentation and returned his customer's call.

Sound familiar? How many times do you find yourself running around, expending energy and not really getting anything done. Customer service representatives, in an effort to follow-up on customer requests, spend some time away from their desk. Customers have come to expect that with one call they can do it all. They expect you to be at your desk, waiting for their call and able to resolve their situation immediately. Unfortunately, customer

---

needs don't always fit into the way your company operates.

Meeting customer expectations is stressful with all the multi-tasking involved with customer service work. Reacting to this stress and pressure is harmful. Allowing these small obstacles to become bothersome will have a cumulative effect on your behavior and may result in more stress which will further impede your communication with customers and your on the job performance.

Interruptions

Linda Morgan processes applications for medical claims. She also handles telephone inquiries regarding medical benefits and the status of patient applications. Reviewing hundreds of applications a day takes so much concentration that when her phone rings, Linda loses her place and has to start from the beginning.

When the phones are really busy, Linda gets upset because the pile of applications to be reviewed grows and grows. Linda keeps thinking to her self, "If only the phones would stop ringing, I'd be able to do my job".

Interruptions are a major cause of stress in customer service and multitasking is a reality in today's business world Handling

Figure 15

Interruptions make our lives more Stressful

How can I get this mail done when people keep calling me!

multiple priorities is stressful because, since you can only do one task at a time, by doing one you abandon the other. (See Figure 15). Then if you are interrupted with another critical task, neither of the previous tasks will get done. For a person who wants to do it all, interruptions and incomplete tasks are frustrating and stressful.

Unfortunately, multitasking is probably here to stay and since that condition won't change, we will have to. An old maxim goes, *If you always do what you've always done, you'll always get what you've always gotten.* It means that if what you are doing isn't working, then try a change. You must learn to accept the conditions that are essentially unchangeable and focus on the conditions that you can change.

Working with your supervisor may help. I find it beneficial to allow CSRs time away from their desks so they can complete tasks without being interrupted. My experience in attempting new strategies usually leads to other resourceful ways of overcoming obstacles. You must keep trying new ideas! Don't just sit there and let interruptions render you ineffective. Reserving time for thinking and planning are important, especially if you plan on succeeding in business.

If a task or tasks are not getting done, due to interruptions, it's easy to become overwhelmed and think you will never get to them. I suggest taking mouse steps! This means breaking down these tasks into smaller ones and completing them one at a time at your convenience.

Ten Strategies for Staying Courteous Under Stress

1. The Rudiments - Your mother probably conducted your first and most important communication skills seminar. She taught you to say "Please" and "Thank You." Believe it or not, these rudiments still work wonders. A little politeness goes a long way. Unfortunately, when stress renders us ineffective, the rudiments are the first thing to go.

2. Identify Stress Signals - Learn to recognize when you are under stress. It is that hyperactive feeling you have after closing a call with an upset customer. You know, that tightening in the chest or feeling like you are out of breath or like you just have to get up and take a walk. The stress mechanism gives you an immediate burst of energy, so you'll want to do something physical, it's only natural.

3. Proper Breathing - Force yourself to breathe slow. This is important. Doing this consciously forces you to think rationally about what just happened. You had an unpleasant event that forced your metabolism to increase. Breathing slowly will help offset the sudden change in your metabolic rate and allow you to remain courteous.

4. Rational Thinking - Thinking rationally about what caused you to get stressed is the first step towards keeping things under control. You are probably not being threatened with bodily harm, yet the stress mechanism makes you feel as though you are. Challenging your own thought process can help you overcome the effects of stress in customer service. Our society is very big on not thinking. We are very big on doing. Action is seen as the first and foremost activity. Thinking is secondary. We feel it's more important to get the next call rather than take a few moments to do a ten second readjustment. Since irrational thinking produces stress, doesn't it make sense that rational thinking will help manage it? Investing a few seconds in rational thinking will pay off with improved performance and increased customer satisfaction.

5. Change your Environment - Take a break when you really need one. When stress renders you ineffective, change your scenery. Find a window, and look for something peaceful like a tree, a flock of birds or a garden. These short, five minute, readjustments will revitalize you and enable you to regroup so you can continue working, more effectively.

---

6. Learn to Forget - Don't replay bad experiences in your head, over and over again. Doing this reduces your self-esteem and drains much needed energy on an activity that yields absolutely no benefit. In addition, it distracts you from the customer you are presently working with and it may cause you to appear discourteous. Substitute rational thinking for this self-destructive behavior. Your mental attitude will affect your behavior, so maintain a positive, can-do outlook on your job.

7. Wet Your Whistle - Keep a liquid refreshment such as fruit juice, water or a soda on your desk. I know this sounds silly, but you will need it. Here's why. One of the results of the stress mechanism is a change in your voice. The voice becomes hoarse or raspy when you become stressed. Since your voice is your primary communication tool, you should ensure it is always operating at peak performance. Doing this will enable you to maintain a calm, consistent and courteous tone of voice.

8. Accentuate the Positive - You will appear more courteous when focusing on a positive outcome instead of a negative one. For example, telling a customer, "This isn't as bad as your last problem" instead of, "This is much better than your last situation," focuses on the down side of a situation. You should direct the customer's attention towards an optimistic future by reinforcing the positive potential of a situation, not the negative possibilities.

9. Don't Blame your Customers, Assist Them - You should be careful not to place blame, even when the customer does something wrong. If after a customer describes a problem, and you qualify that the problem is a result of something the customer did, and you say, "You're wrong! It doesn't work that way.", instead of "I may be wrong, but I believe it operates a little differently. Please allow me to help.", the problem can be exacerbated. Words are weapons. The words you use, during these events, will make you appear either courteous or antagonistic. Substituting the word "I" for "You" takes the emphasis (and blame) off the customer and neutralizes the event.

10. Do Not Retaliate - Use restraint not retaliation. Some customer's only concern is satisfaction, which means getting aggressive to get what they want. When you are working with rude or abusive customers, and you feel your stress mechanism activating, use restraint instead of retaliation. An angry customer's behavior is unpleasant, but if you feel the temptation to retaliate, don't. This will only escalate a situation, not contain it. Behave like a professional. Speak in a calm consistent tone and help the customer conform to your professional behavior.

If you retaliate, one of two things will happen. (1) The customer may counterattack more severely and the situation will escalate out of control. This will make your stress even worse as the customer becomes more difficult to work with. (2) You will damage the customer's self-esteem and force him to submit for the short term, but your company will lose the customer for the long term when he purchases some place else the next time.

Maintain A Resilient Personality

Many people go to great lengths to keep stress and adversity in their lives at a minimum. They want to remain in a state of homeostasis. This term is derived from the Latin *homeo* (same) and *stasis* (lack of motion). Experts believe that living this way is unwise and unrealistic. Insulating ourselves from stressful events is impossible. Stressful events are unavoidable and, by their nature will periodically shake us up, disrupt our lives and force us to grow. The sad part about avoidance is it feels good, temporarily. So good, that the temptation to repeat it becomes stronger and stronger. This type of behavior can become a phobia.

Each stressful experience you overcome prepares you to handle the stressful situations that lie ahead (see Figure 16). As these

Figure 16

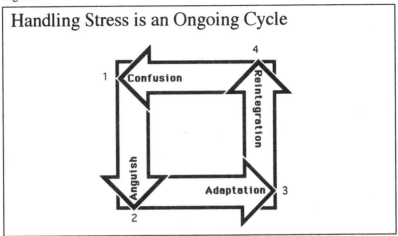

Each stressful experience you overcome prepares you to handle the stressful situations that lie ahead.

experiences leave us in a temporary state of emotional anguish and confusion, they also instigate adaptation and reintegration, if you have a resilient personality. As you overcome stressful experiences you become more knowledgeable and adaptive. This will allow you to control events, instead of events controlling you. This will build your confidence and add positive/manageable experiences to your memory bank to rely on for tomorrow.

This cycle of *confusion, anguish, adaptation* and *reintegration* is a continuing process. As you improve, the difficult tasks you once thought were so unbearable become routine. In life we learn by experience. I encourage you to confront adversity head on. Only by building a reservoir of positive outcomes to negative experiences will you grow in your personal and professional life. Hope for a positive future is based on your past performance and it affects how we work in the present.

Chapter 8 — Key Points:

- Think rationally and focus on the root cause of problems.
- There is no cure for stress, but it can be handled.
- A certain amount of stress is positive and productive.
- Be aware of your body's reaction to stress.
- Don't allow stressors from within your company impede your performance.
- Stay courteous, even when under stress.
- Each stressful experience you ocvercome prepares you to handle the stressful experiences that lie ahead.

Chapter 9:
Changes in Customer's Behavior

Why Customers Get Upset

Jack Smith took a long time to make up his mind. He always did research before he bought anything. When he finally bought what he wanted, he was proud of his purchase. But, after a few months, it began to malfunction. Jack was disappointed.

After explaining the problem to his dealer, Jack was told to bring his unit back for evaluation. The salesman qualified that Jack's unit needed service. He explained that they had an on-site service department that would be able to help. In about a week, the dealer called Jack to inform him that his unit was ready to be picked up. A few days later, Jack's unit malfunctioned again. Jack was frustrated! This was not fair. He spent good money for something that wasn't working.

He contacted his dealer again. The salesperson explained to Jack that his unit was very complex, and although their service department was good, it might be better to return it directly to the manufacturer for repair. After all, they designed and manufactured it. Surely, they will know how to fix it.

In about two weeks, Jack received his unit back from the manufacturer. Everything seemed fine. Jack couldn't wait for the weekend so he could use it. But, on Saturday morning, Jack's unit malfunctioned again. That was it! Now he was angry. This was war.

Jack couldn't wait for Monday morning. He was going to call that company and speak to someone in charge. If that company was going to provide inferior products and inconvenience their customers, then they were going to bear the consequences. Jack thought about what he would say on Monday. He decided to insist on compensation for his inconvenience.

He would demand a newer updated model that the manufacturer had just advertised. It had nicer features and it cost a little more, but Jack didn't plan on paying anything for it. He wanted it for free, because they owed him that much. And if they wouldn't do it, Jack would threaten to sue.

On Monday morning, the service manager at the manufacturer received a call from an irate customer, named Jack Smith. The call lasted about 5 or 10 minutes. Soon both Jack and the service manager were speaking amicably.

The service manager convinced Jack that the unit would need to be returned to the company again. Jack agreed, and after that repair the unit worked fine. Jack bought other products made by this company because he knew if there were any problems, he knew who to call.

A negative experience with one of your company's products or services may change a customer's behavior from good to bad. Customers like Jack Smith, start out being patient when something goes wrong. However, when his minor inconvenience became a major catastrophe due to a breakdown in the manufacturer's in-house system, Jack's behavior changed for the worse. These customers are still nice people, it is the inconvenience or embarrassment of the problem that makes them behave otherwise.

Their conversation is confrontational because they feel as though they lost something and they want to be vindicated. This includes winning back some lost ground, so they aim for a win/lose result: they win and you lose. These perceptions, whether they are right or wrong, are real to a frustrated customer. Customers' past experiences make up the sum of who they are and how they behave. Our job is to remove any resemblance of a conflict and aim for a win/win result.

I once worked with an abrasive customer who believed the only way to get what he wanted was to sue. His reason for behaving this way was that he once had a difficult experience with an insurance company that, after his wife passed away, refused to pay the death benefit. Retaining a lawyer and threatening the insurance company with a potential lawsuit motivated them to pay him. Since that event, he became a customer to be reckoned with. He mistrusted corporations and threatened to sue whenever he didn't get what he wanted.

He had a difficulty with one my company's products, called our customer service department and demanded compensation that was unreasonable. He also threatened to sue if we didn't satisfy his demand. Although we didn't yet know about his dilemma with the insurance company, this customer's behavior was affected by his past experience.

Calls involving legal matters are passed up to me by our CSRs. I listened to his complaint regarding our product and offered to repair it under our company's warranty. He declined my proposal and demanded more than warranty service. He wanted his product replaced with a newer, more expensive model. I told him that although I could not fulfill his request, I could solve the problem immediately and at no cost. He declined my offer and we were at an impasse. He said he wanted to consult his lawyer about his next move, so we agreed to contact each other the following day at an appointed time to discuss the matter further.

The next day I phoned him and instead of getting right down to business, I engaged him in friendly conversation for a few minutes. I told him about my background and he told me a little about himself which included his experience with his wife's insurance company. I listened as though I were a close friend, providing consolation and support as he described the details. When the conversation turned to business and I brought up the issue of his defective product, he agreed to let me fix the product for him without any mention of a lawyer or unreasonable demands. I was shocked at his change in behavior (but I didn't argue with him). Why did he change his mind? Because my relationship changed from adversary to ally. I believe his mistrust of corporations forced him to respond with unreasonable demands. Then, by allowing time for him to cool off and think rationally about the situation, then start anew, I overcame this obstacle. Never underestimate the power of friendly, personal, one-on-one interaction.

Rising Expectations

During the last decade, I have witnessed a significant change in customer behavior. Customers are simply more demanding than ever! Customers' demands for service are rising for many reasons. The first reason is that many customers are receiving quality service from some of the companies they buy from. Having received quality service one or more times, customers begin to expect all the companies they buy from to meet the same standards.

If customers are used to being serviced well by company X, why should they have to settle for anything less from company Y? Quite possibly, the CSRs at company Y will recognize a shift in their customer's attitudes, not because the CSRs are ineffective, but because their customers' expectations have been heightened by the competition. Companies are driven to quality service by the loss of customers to the competition. Failure to measure up to customer expectations will eventually result in customers going elsewhere.

Evidence of the recent changes in customer behavior are illustrated in Figure 17. It indicates how customers defined quality in 1985 and in 1990. This chart is a little dated, but this trend still holds true. In 1985, customers were more concerned about the brand name of an item rather than whether the item was reliable. Years ago, customers felt a sense of security when buying merchandise from well-known companies. Since service and reliability among many companies were similar, buying from a well-known established company ensured long term support. However, the companies that improved their service and reliability began to stand out among the competition, and customers noticed the difference. The maxim, *A bigger company is better*, is now replaced with *A better company is better*.

These recent changes in customer preference are consistent with recent findings: "...eight in ten executives report their

Figure 17

How is Quality Defined?

How Americans rank Quality's Components in 1985 and 1990

	1985		1990
1	Well known brand name	1	Reliability
2	Workmanship	2	Durability
3	Price	3	Easy Maintenance

Source: Fortune Magazine

---

[3] The State of Quality Customer Service in America:1990, John Hancock Financial Services ©,

company's commitment to quality service has risen."[3] This increased commitment for quality service is the reason for rising customer expectations. Companies that are successfully improving their service and reliability will continue to attract new customers, further widening the gap between them and their competition.

Another reason for the recent changes in customer behavior is that many are paying more for almost everything and therefore expect more from the companies they buy from. Customers assume more expensive products will perform more reliably than less expensive products.

Customers' demands for better quality are something they are willing to pay for. They don't mind paying more for reliability because customers don't have time for interruptions in their busy lifestyles.

Americans are working more hours than ever. During the last two decades, the continuing decline of real wages coupled with the increasing costs of health care and housing, have forced Americans to work more hours. In fact, we work about 150 more hours a year. During the 1980's, paid time off such as vacations, holidays, sick leave and personal days fell roughly about 15 percent. The number of people who say they have less free time than they did five years ago is twice as high as the number who say they have more. Even more startling is the revelation that many employees would be willing to give up a day's pay to have an extra day off from work. So if a customer's car breaks down or if something else goes wrong, they have less free time to deal with this inconvenience. The CSR who handles a phone call from one these customers might have to deal with the resulting anger and frustration.

With the migration from the urban city to the suburbs, people also spend more time getting to and from work. Figuring in the rise in work hours, commuting time and the decline in days off, Americans are spending about an extra month working each year.

People spend more time working now than they did 20 years ago. Why? Because they need the money. Why? Because they want the best. Why? Because they believe best is more reliable. There is

Figure 18

Rising Expectations

Customer expect more because they know they can get it. If your company can't deliver, customers will find a company who can. Dealing with rising expectations takes its toll.

no doubt that the squeeze for more time and better quality has caused a dramatic shift in attitudes.

One more reason for the change in customer's behavior is that they are now more aware of their rights and power as consumers. In recent years; product tampering, disease causing fish and poultry, product recalls and other health and safety risks have led to a general mistrust of corporations. Customers protect themselves from harm because they don't feel the government can do it for them.

Some customers are investigating the track record of companies before buying from them. The Better Business Bureau reports that pre-purchase inquiries are up 50%, from about six million contacts in 1985 to more than nine million contacts in 1992 (see Figure 19). The Better Business Bureau will issue, upon request, a performance report on a company's good-faith response to valid customer complaints.

Figure 19

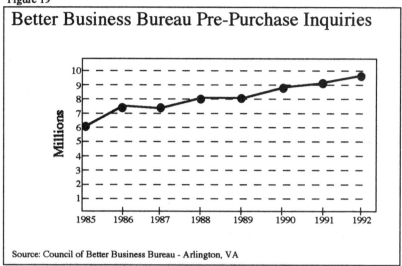

Better Business Bureau Pre-Purchase Inquiries

Source: Council of Better Business Bureau - Arlington, VA

Have you noticed changes in your customer's behavior during the last few years? If so, rising customer expectations, busier lifestyles, the demand for quality and a mistrust of corporations are the reasons. I predict that customers' expectations will continue to rise and make customers more likely to complain if things are not right. Hence the need for good communication skills in customer service.

Chapter 9 — Key Points:

- Customers expect more from the companies they buy from.
- Customer expectations are being heightened by the companies that provide excellent service.
- Customers are now more aware of their rights and power as consumers.
- Customers have less time for interruptions in their busy lifestyles.
- Customers are paying more for almost everything and therefore expect increased reliability.
- Pre-purchase inquiries with the Better Business Bureau has been increasing steadily during the last eight years.

Chapter 10:
Customer Service Technology

Excellent communication skills with efficient, user friendly technology is a winning combination. Technology enables a customer service organization to make the most of its resources. I have worked extensively with each of the technologies described in this chapter. Applying these technologies to my department has had a positive difference on how I manage and conduct customer service.

The following story illustrates how some companies might view new technology. Don't let it happen to you.

Mr. Greene, a vice president of customer service operations at the ABC Insurance Company is a hard worker, but he's not very computer literate. He delegates most computer related projects to Mr. Hunter, the customer service manager. Each day, hundreds of policy holders call ABC's customer service department to request a current policy statement. ABC used to mail the statements, until Mr. Hunter received approval for a few hundred dollars to buy a fax machine. Since a fax costs

Figure 20

Cost Is Not The Only Factor

It's too much money! Don't waste my time with this nonsense. Can't you see we have all this faxing to do?

less than a postage stamp and some of their customers have fax machines, Mr. Hunter assigned one person to manually fax statements to customers. This solution worked for awhile.

Then Mr. Hunter learned about fax on demand. He believed that purchasing a fax on demand system would be a better way to service customers. The system would be available 7 days a week, 24 hours a day and it would eliminate the need for one of their CSRs to manually send faxes. In addition, Mr. Hunter could include marketing and sales documents regarding new products provided by ABC. He contacted a sales representative from the ACME Telephony Company.

When Mr. Hunter introduced the sales representative to his boss, all Mr. Greene wanted to know was "How much will it cost?" The system was substantially more expensive than a stand-alone fax machine, and Mr. Greene thought it was too much money. Mr. Hunter was reprimanded and told to get back to work. Meanwhile, *the ABC insurance company's cost of doing business keeps increasing. ABC passes these cost increases onto its customers. Eventually more and more of ABC's customers are approached by the competition, who have implemented newer technologies and can offer better services for less money. As a result, ABC keeps losing market share and its operating expenditures keep rising.*

The technology in the story could have been almost anything. The point is that cost should not be the only factor by which new technology is judged.

The Cost Of Technology

Technology costs money, but don't let this scare you. Technology that is carefully planned and implemented will pay for itself, sometimes in a matter of months. Let's consider ABC Insurance Company's situation over a four year period (see Figure 21). In year one, ABC employs 8 CSRs. Each CSR makes an average of $29,000 annually. The toll-free phone bill for ABC is about $2000.00 a month. The annual operating expenditures for people and incoming calls is about $250,000.00. If ABC's incoming calls are growing at a rate of about 20% annually, additional CSRs will need to be hired.

To manage the growth of incoming calls, ABC will need to hire two CSRs for each of the following three years. During this time, their expenditures for salaries will increase by $174,000.00. That's a lot of money!

Now, let's say that in December of the first year, ABC decided to buy a fax on demand system for a purchase price of $16,000.00. This means that their second year starts off with the benefit of this new technology, which is capable of handling about 20% of customer's requests. The initial investment of $16,000.00 for the fax on demand system pays for itself in the first few months of the second year.

Figure 21

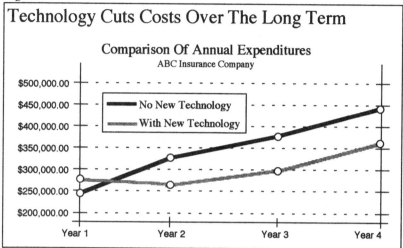

Technology Cuts Costs Over The Long Term

Comparison Of Annual Expenditures
ABC Insurance Company

As the above graph indicates, new technology purchases will increase expenditures for the short term only. This new technology will reduce operating costs over the following three years.

Over the same four year period, even with incoming calls increasing at 20% annually, the salary expenditures increase by only $87,000.00, which is 50% less than if no new technology was purchased. The message is clear. Technology not only pays for itself, it saves money over the long term.

Fax On Demand

As its name implies, fax on demand allows customers to automatically retrieve faxed documents on demand. If your customers have a fax machine, or a fax modem, they can call your fax on demand system, request a catalog of your documents, then call back and order specific documents that interest them. You can create documents that answer the questions routinely asked by customers. Or you can make available product spec sheets, price lists and news releases. The applications are endless. Customers can retrieve these documents 24 hours a day, seven days a week, without taking up a CSRs time.

There are more fax machines out there than you think! In 1986 only .26% of Americans had the ability to fax from their homes. In 1994, it was 17.5%. There's a good chance that many of your customers either own, or have access to a fax machine. Eric Arnum, editor of the *Electronic Mail and Microsystems Newsletter*, reports that there are about 24 million fax machines worldwide. He also indicates that sales of fax modems are about 10 million units annually.

Each fax on demand system has incoming and outgoing lines. The incoming lines handle the incoming calls and the outgoing lines handle the outgoing faxes. You can get a one-in/one-out system, a two-in/two out system, four-in/four-out system and on and on, depending on your traffic.

Fax on demand systems come in two varieties: *one-call* and *two-call* (see Figure 22). In a one-call fax on demand system, the caller phones into the system from a fax machine and requests the documents he needs. The fax on demand system locates the documents and tells him to wait for a tone. When he hears the tone he can press "Start" on his fax machine and documents emerge.

In a two-call setup, after the caller calls in and requests documents, he enters in a fax number using his telephone keypad, and hangs up. The system then calls the number entered by the caller and delivers the documents immediately. Systems can be configured so that the faxes are scheduled to be sent at a later time, such as late at night, when the long distance costs are down.

Companies looking to keep costs down like the one-call system better because phone costs are on the caller's nickel. Customers prefer the two-call system because it's more convenient. Customers can use it from any touch tone phone and they don't have to be at a fax machine to request a fax. When you consider how precious time and convenience are these days, more customers will use the two-call system. So, if you're looking to attract customers, use the two-call system.

Making your fax on demand system logical and friendly is critical to its success. Publicize your fax on demand system often.

Figure 22

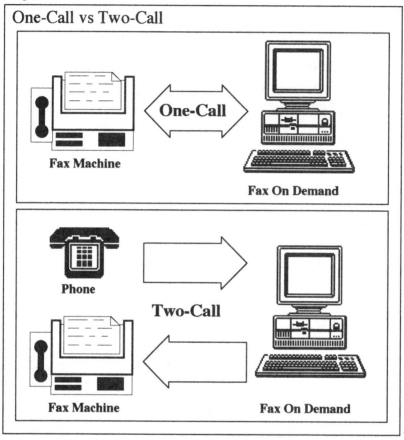

Send out press releases, include your fax on demand system phone number on all your company's letterhead, business cards and brochures. The more your customer's use it, the more you'll save, because they won't be calling your CSRs.

Most fax on demand systems will capture the time and date of transactions in addition to your customer's fax machine number. This information enables you to audit your system and measure the popularity of your company's different documents. No matter what type of business you're in, fax on demand can help. Whether you're looking to handle only 25 requests a day or 2,500 requests a day, there's a system for you.

Automated Call Distributor

An automated call distributor (ACD) is used for routing incoming calls to their destinations. The destination is usually an extension number. When you call into an ACD, the automated attendant usually greets you and asks for an extension number. If you know your party's extension number, you may dial it on your touch tone keypad. If you don't know your party's extension, the system usually defaults to a live operator that can route you to your party's telephone. A feature that will relieve your operator of transferring calls is a "spell-by-last-name" directory. This feature allows callers to type in the first three letters of their party's last name. This enables callers to get through to their party without knowing an extension number. ACD's enable customers, who know their party's extension, to get to their destination faster, without having to go through a live operator.

When setting up an ACD, it's best to make a flowchart of how you believe your system should be set up. Use Figure 23 as a working example. Making the flowchart will enable you to get visual confirmation of your system's logical sequence and user friendliness. Limit the number of choices you offer callers in a single menu. More than four or five choices may confuse your callers. Don't force them to wade through a zillion menus before they reach the choice they want.

Make it easy for your callers to get to a live operator. Use the "0" for operator in your system. If a caller selects "0" because their party is on the phone, enable the operator to get through to that party to let them know about it. Give all phones second lines. Doing this will make the operator feel more helpful.

The XYZ computer company developed a truly innovative computer product; The Widgetizer. This product took the consumer market by storm. Almost immediately, the calls started coming in by the thousands. Customers called with technical and application

Figure 23

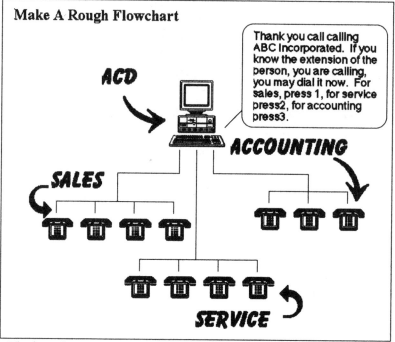

Make A Rough Flowchart

Thank you call calling ABC Incorporated. If you know the extension of the person, you are calling, you may dial it now. For sales, press 1, for service press2, for accounting press3.

ACD

ACCOUNTING

SALES

SERVICE

The best place to start designing an ACD is on a sketchpad.

questions and the customer service manager couldn't hire CSRs fast enough.

Most of the technical questions were related to some minor software bugs. The CSRs were always able to suggest workarounds for customers, but this process took time, which made it an expensive procedure. Finally, a new version of software was released and the customer service manager had a great idea. He knew that thousands were going to learn about the new operating system. Customers would want to know how they could get it, its new features, what bugs it fixed and what it cost.

Anticipating this influx of calls, he changed the company's voice mail greeting. The new greeting mentioned the new software and suggested an option for customers to learn more about it. When customers selected this option, they heard a message telling them *how they can get it, what new features it had, what bugs it fixed and what it cost. This option was available 24 hours a day and seven days a week. He saved his company thousands of dollars in telephone time by giving customers information when they needed it.*

Computer Telephony Integration

No matter what type of business you're in, your operation will benefit with an effective call center implementation. Whether you're a kitchen appliance manufacturer, a food processing plant or a textile mill, your customers, your vendors and your end users will need to contact you. Many of these interactions will include routine questions or inquiries. The call center technology that can handle this is inexpensive and easy to configure. For example, designing an FAQ (frequently asked questions) kiosk on your ACD (automatic call distributor) or voice mail server can help provide better value with a 24 hour 7 day a week solution.

Only a couple of years ago, very sophisticated and expensive call center technology, such as ANI (automatic number identification) and DNIS (dialed number identification service) enabled software and hardware, was only available to larger corporations with huge budgets for capital spending. Today, computer telephony integration (CTI) allows both big and small companies to enjoy the same sophisticated features. Some small, feature rich CTI solutions cost as little as $2000.

ANI, for example, can give your call center the ability to integrate your computer and telecom systems to deliver a customer's phone call along with the caller's pertinent data. Thus displaying his name, account number, address, buying trends, etc. right to the CSR's computer screen. This eliminates the time a CSR would normally take to ask for the customer's name and account number and then look up the records. When you consider the cumulative time saving, over thousands of calls, this technology alone will save you big money in 800 line charges and CSR labor.

Let's take this example one step further. Suppose this customer has a problem that requires a specialist. In this case, the CSR simply presses a button on his computer screen and transfers the call and pertinent data to a second level specialist. The specialist receives the phone call and the caller's data on his computer screen saving the customer the trouble of having to repeat his saga to a second person. CTI applications like these please customers because they save time and add value.

Does CTI sound too complicated? It's not. Today's applications are user friendly with a graphical user interface that allows almost anyone to configure and set up these systems.

Voice Mail

Voice mail is a vital part of customer service in today's business world. This technology is an extension of the stand-alone answering machine. Instead of tape, voice mail systems use digital memory to store callers messages. These systems are much more flexible and full-featured than answering machines.

In a study, conducted by the Voice Messaging Educational Committee (VMEC), more than 89% of the companies that use voice mail found it to be an important medium for business communication. The VMEC study also revealed that 78% of these companies reported that voice mail improved their job productivity.

While we agree that voice mail is a vital link in today's business communications, we might also admit that voice mail can sometimes be a nuisance. This is especially true when customers feel compelled to speak with a live person. My own experience has taught me that customers will use voice mail confidently, when they believe, based on past experience, that their call will be returned promptly. Customers become hesitant about leaving voice mail messages because of past, unpleasant experiences involving long delays in call backs. This reality places the burden of responsibility on the person owning the voice mail box.

Remember those "While You Were Out" message slips? I used

Figure 24

Which Would You Rather Use?

Hi John. I have great news! We've got the Johnson account. It all hinges on us getting them a proposal tonight. There's a file on my PC named JNSON.DOC, fax it to 212-555-1234. Call me with any questions at 415-555-1234.

Date: 5/25/95   Hour: 1:15
To: Steve

While You Were Out

M   John
Of: Sales Department
Phone: 415-555-1234
          Area Code   Phone Number

Telephoned	✓	Returned Call		Left Package	
Please Call	✓	Was In		Please See Me	
Will Call Again		Will Return		Important	✓

Message
Call him. It's important!

Voice Mail   **Message Slip**

A person's tone of voice says much about their current state of mind. You'll be able to sense this with voice mail, not so with message slips.

to be a slave to them. I'd come back from lunch and find a stack of these slips on my desk, which included a name, a phone number and a cryptic message. The information provided on these slips didn't allow me to do much preparation before returning a call. I knew who to call, but I still had to learn why. So I'd call and find out what my customer needed, then ask them to hold while I did research or offer to call them back. It was a waste of time.

Voice mail resolved these situations for me. With voice mail, I not only get the name and phone number of the person who called, I also get a detailed message. A detailed message of this kind would be impossible to write on a slip. There are too many words. With voice mail, however, I get all the information I need. Having all the detail enables me to prepare before calling. I can satisfy my customer's request with just one call, instead of two.

I also get to hear my customer's tone of voice. A person's tone of voice says much about their personality, aggressiveness and current state of mind. Knowing in advance whether my customer is friendly, relaxed, anxious or irate gives me an advantage. In addition, voice mail systems keep track of the time and date of each message.

Voice mail can be forwarded to someone else within your organization along with a personal introduction from you. You can also retrieve your voice mail messages while you're away from your desk. This is a convenient feature, especially when you're on the road traveling.

E-Mail

E-mail is an abbreviation for "electronic mail." While E-mail can include all the on-line services available worldwide, I am only referring to it's use within an organization. It usually runs on a computer network so many users can stay in touch.

E-mail will enable you and your CSRs to share information without sending memos and other random pieces of paper around the office. If you manage a department of 10 or more CSRs, this can be a tremendous advantage. Imagine if you had to walk around to each cubicle and hand out a memo or interrupt each CSR whenever a late breaking change had to be distributed. In fact, I didn't have to imagine this, I used to do it! What a waste of time.

When someone sends you e-mail, the message is stored in your computer's in-box. Incoming e-mail messages are displayed on a user friendly computer screen along with the date of e-mail and a short description of its contents. Seeing all your e-mail at a glance allows you to select which messages need to be read first. This is something you can't do with a paper in-box.

Another advantage of e-mail is that you can write one message and have it distributed to hundreds on people at the touch of a button. This is because e-mail allows you to create distribution

Figure 25

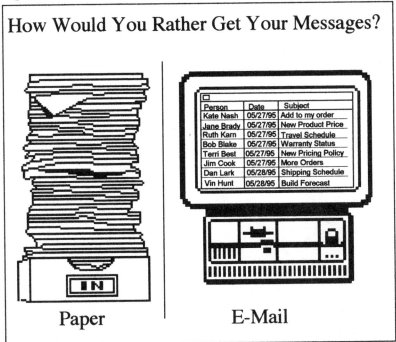

How Would You Rather Get Your Messages?

Person	Date	Subject
Kate Nash	05/27/95	Add to my order
Jane Brady	05/27/95	New Product Price
Ruth Karn	05/27/95	Travel Schedule
Bob Blake	05/27/95	Warranty Status
Terri Best	05/27/95	New Pricing Policy
Jim Cook	05/27/95	More Orders
Dan Lark	05/28/95	Shipping Schedule
Vin Hunt	05/28/95	Build Forecast

Paper   E-Mail

Paper memos only get covered up. E-mail is neatly displayed on a computer screen allowing you to view the sender, date, and subject matter of each message.

lists of people who work in your organization. You can have one distribution list for the engineering department and another list for accounting.

In addition, many e-mail products allow you to attach a file. This enables you to send one person, or a group of people, a spreadsheet or a document along with your e-mail message. Once you start using e-mail, you'll wonder how you ever got along without it.

Training is an important element of an e-mail installation. All e-mail users must be trained to exploit the system's features. Don't assume that people will share information electronically simply because the system has been installed.

Figure 26

It's easy for a customer service manager, or anyone else, to get seduced by new technology. It's so seductive that you might just find yourself spending more time with the technology than you do with your people. Don't let this happen to you.

Don't Get Seduced

A computer, once it's set up and configured will perform redundant tasks, without much maintenance. That's the big benefit. You invest the time up front and then let it do what it does best. Not so with people!

Technology is a tool. Treat it as such. If your systems are set up well, more of your time should be freed up to work with your

CSRs. This is an opportunity, so don't miss it. Your CSRs will need your appraisals, encouragement, feedback and direction. Don't ignore your most important assets - your employees.

When it comes to people, the cost of training is less than the cost of ignorance. If you're a manager, you must make your people the best they can be. Doing this is not easy. Each time new technology is introduced, managers will find themselves in the middle of a training nightmare. Most of this training is about non-stop operational questions regarding the new systems.

Being an effective manager in the midst of all this change takes someone with a clear vision and a resilient personality. A person who plans his work then works his plan.

Chapter 10 — Key Points:

- Technology saves money over the long term.
- Fax on demand is an effective information fulfillment technology.
- Use an Automated Call Distributor to route calls throughout your facility..
- Computer Telephony Integreation products are within the reach almost any company's budget .
- Don't get seduced by new technology.
- Technology is a tool, use it as such.

Chapter 11:
Where has Customer Service Been and Where is it Going

Readers who are new to the customer service industry might enjoy a historical perspective on where this industry has been and where it is going. In the 1950's and 1960's, the key players in a sales and service organization were the field sales representatives. They traveled their territories and met customers face to face while inside CSRs worked the telephones making appointments for field salespeople, handled clerical duties and answered customer inquiries.

CSRs were held in low regard as professionals go. For people looking to move ahead, a job in customer service was often a stepping stone to something else, usually a sales position. Why not! The compensation plans used by sales organizations emphasized growth and new business. The stars of sales organizations were the ones who brought in new accounts. There were no rewards for satisfying existing ones. Since CSRs were not motivated to excel, talented business people avoided a career in customer service and for a while the quality of customer service in America reflected this.

By the mid 1970's, advancements in telecommunication technology made telephones the preferred sales and marketing medium. Companies looking to reduce sales costs allowed their in-house customer service department to sell in addition to providing service. These companies realized that field sales reps incurred the same travel and lodging expenses whether they visited an "A" account that generated $10,000 annually or a "B" or "C" account that generated only $1,000 a year. From a cost/benefit viewpoint, it made sense for CSRs to handle the needs of "B" and "C" accounts between less frequent visits from field sales reps. This would permit field sales reps to support and visit the "A" accounts more frequently. CSRs soon found themselves doing more "real business", in addition to traditional clerical duties.

Customer service inevitably became the "buzz word" of the 1980's. Corporations providing excellent customer service made an additional service investment when handling customer complaints. This investment included toll free numbers, training programs for CSRs and prompt personal service. For companies with vision, this investment eventually translated into long term profit as formerly unhappy customers told others about the excellent service they had received.

As customers were attracted to the companies delivering quality service, the competition realizing losses in revenue and market share stepped up their quality service initiatives. More than any other factor, it is competition that drives improvements in customer service.

Now many corporations require CSRs to have a college degree and sales experience. Gone are the stereotypical receptionist job functions. CSRs are professionals and they will be a major force in the 1990's. The number of telephone sales representatives has risen fourfold since 1984. U.S. expenditures for telemarketing have shot up to $60 billion from $1 billion annually during the past ten years. People are doing more business via the telephone than ever before. The telecommunication industry is booming and will continue to boom as major U.S. cities add new area codes to handle the ever growing need for more telephone lines. Computer technology will continue to make it possible for companies to simplify their customer service operations and provide greater customer satisfaction.

With all the advancements in telecommunication technology, one thing has not changed; the problem solving process still requires personal intervention. When customers get emotional, only another person with a caring attitude and genuine concern will suffice as the problem solver. They haven't invented a computer that can empathize with and console customers. Fortunately, humans are still the ultimate problem solvers.

Technological advancements are happening fast. The magnitude of change, especially in computer-based technology, is out pacing most of our customer's comprehension. The gap between what customers know and the way things really are is widening. This condition frustrates some customers merely because they don't understand. Customer service professionals

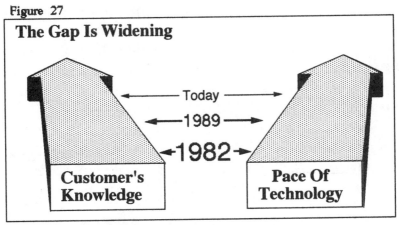

Figure 27

The gap between customer's knowledge and the pace of technology is widening. This trend often manifests itself as difficult behavior.

must be sensitive to this situation and using the strategies in this book will surely help.

Another critical trend that will affect how future customer service will be conducted is the aptitude of entry level CSRs. I conducted a survey involving more than 7000 CSRs among sixty six of America's largest corporations. When customer service managers were asked if entry level CSRs had the necessary skills to do their job, 75% said no. These managers indicated that entry level CSRs did not have good communication skills such as listening, problem solving, telephone etiquette, patience, grammar, common courtesy and empathy.

Human resource professionals will share the burden of this problem as more emphasis is placed on screening potential employees. Only candidates with the best skills will be offered positions. This trend may indicate a problem for the future of customer service as employers will need to invest more time and resources teaching people basic communication skills.

Conclusion

How you use the information in this book is up to you. I provided my knowledge and experience, now you must apply your effort and start using these techniques immediately. As your problem solving skills develop and improve, you will grow in confidence and speak with the voice of authority. Customer service is a challenging industry that demands flexibility from those who provide services because of the constant changes and improvements.

Remember, the winning formula for successful problem solving is: respond immediately; tell the truth; use the contain, qualify and correct method; handle stress by maintaining a resilient personality and the golden rule — treat others the way you want to be treated.

Good luck!

Index

A

Abbreviations, 80
Abusive Customers, 30
Accurate Answers, 36
Adaptation, 101
Adverse Situations, 21
Aerobics, 94
Aggressive Customers, 30
Absenteeism, 90
ANI, 121
Anger, 26, 39, 81
Antagonize, 33
Anxiety, 94
Arnum, Eric, 117
Attract New Customers, 117
Attitude
   courtesy, 97
   positive, 22
   problem solving, 38
   rational thinking, 26, 49
   stress, 89

Automated Call Distributor, 119
Automatic Number Identification, 121

B

Better Business Bureau, 85, 110
Blame, 21
Blood Pressure, 92
Busier Lifestyles, 110
Business Relationships, 30

C

Claims, 85
Close Ended Questions, 56
Closing Phone Calls, 32
Commitment, 108
Commuting, 108
Complaint letters, 85
Concern, 15, 131

S

Satisfy, 16, 19
Service Investment, 130
Solving Problems
   correct attitiude, 38, 39
   e-mail, 77
   lack of skills, 36
   offensive action, 47
   origin of, 39
Speech
   affects of stress, 92
   authoritative, 49, 63
   calm consistent tone, 30
   confidence, 64, 101
   consonants, 31
   control conversation, 36
   40, 48
   proper closing, 32
   proper greeting, 31
   restraint, 28
   uhm, 31
   vocal inflections, 78
   vowels, 31
Stress, 89
   affects of, 90
   cure for, 94
   cycle of, 101
   harmful affects, 94
   instigation of, 90
   interruptions, 96
   metabolism, 91, 93
   positive effects, 92
   react to, 94
   response to, 92
   staying courteous, 97
   unavoidable situations, 100
   voice changes, 92
Subconscious, 94
Supervision, 37, 49
Survey, 36, 90, 108, 132

T

Technology, 113
   as a tool, 126
   ANI, 121
   CTI, 121, 122
   DNIS, 121
   cost of, 115
   cost effective, 121
   inexpensive, 121
   pace of, 131
   savings of, 121
   seduced by, 125
Telecommunication, 131
Threats, 37, 49, 82
Time Stamp
   fax on demand, 118
   voice mail, 120, 121, 122
Training, 36, 125
Truth, 15, 16, 19
Two-Call, 117

U

Uhm, 31
Unreasonable Demands, 49, 58, 82
Unacceptable Behavior, 38, 39

V

W

PUBLISHED BY
TELECOM BOOKS

Telecom Books publishes books and magazines and organizes trade conferences on telephones, telecommunications, local area networks, PC data processing, desktop office automation, data communications and sales automation software and hardware. It also distributes the books of other publishers, making it the "central source" for all the above materials. Call or write for your FREE catalog.

OTHER BOOKS BY TELECOM LIBRARY

- Newton's TELECOM Dictionary
- The Dictionary of Sales and Marketing Technology Terms
- TELE-Stress
- Negotiating Telecommunications Contracts
- Buy Short Haul Microwave
- The Perfect Proposal
- The Inbound Telephone Call Center
- Student Communications Services

- The Complete Traffic Engineering Handbook
- SONET: Planning, Installing & Maintaining Broadband Networks
- The Guide to Frame Relay and Packet Networking
- Frames, Packet and Cells in Broadband Networking
- 137 Money-Saving Secrets Your Phone Company Won't Tell You
- Which Phone System Should I Buy?

FREE CATALOG OF BOOK

The Telecom Books publishes books itself, and also distributes the books of every other telecommunications publisher. You may receive your FREE copy of our latest catalog by calling 212-691-8215, or by dropping a line to Christine Kern, Telecom Library Manager, at the address below. You may order your Telecom Library books by calling 1-800-LIBRARY, or fax your order to 212-691-1191.

QUANTITY PURCHASES

If you wish to purchase this book, or any others, in quantity, please contact:
Christine Kern, Manager
Telecom Books, Inc.
12 West 21 Street
New York, NY 10010
1-800-LIBRARY or 212-691-8215   Facsimile orders: 212-691-1191

TELECOM BOOKS, INC. 12 WEST 21 STREET, NEW YORK, NY 10010
212-691-8215   1-800-LIBRARY